Sustainable Development Goals Series

The **Sustainable Development Goals Series** is Springer Nature's inaugural cross-imprint book series that addresses and supports the United Nations' seventeen Sustainable Development Goals. The series fosters comprehensive research focused on these global targets and endeavours to address some of society's greatest grand challenges. The SDGs are inherently multidisciplinary, and they bring people working across different fields together and working towards a common goal. In this spirit, the Sustainable Development Goals series is the first at Springer Nature to publish books under both the Springer and Palgrave Macmillan imprints, bringing the strengths of our imprints together.

The Sustainable Development Goals Series is organized into eighteen subseries: one subseries based around each of the seventeen respective Sustainable Development Goals, and an eighteenth subseries, "Connecting the Goals," which serves as a home for volumes addressing multiple goals or studying the SDGs as a whole. Each subseries is guided by an expert Subseries Advisor with years or decades of experience studying and addressing core components of their respective Goal.

The SDG Series has a remit as broad as the SDGs themselves, and contributions are welcome from scientists, academics, policymakers, and researchers working in fields related to any of the seventeen goals. If you are interested in contributing a monograph or curated volume to the series, please contact the Publishers: Zachary Romano [Springer; zachary.romano@springer.com] and Rachael Ballard [Palgrave Macmillan; rachael.ballard@palgrave.com].

Melissa Liow Li Sa · Sam Choon-Yin

Sustainable Urban Development in Singapore

Imagining Walkability in an Urban Concrete Jungle

palgrave
macmillan

Melissa Liow Li Sa
PSB Academy
Singapore, Singapore

Sam Choon-Yin
Kaplan Higher Education Institute
Singapore, Singapore

ISSN 2523-3084 ISSN 2523-3092 (electronic)
Sustainable Development Goals Series
ISBN 978-981-99-5450-6 ISBN 978-981-99-5451-3 (eBook)
https://doi.org/10.1007/978-981-99-5451-3

Cover illustration: Francois Roux/Alamy Stock Photo

This Palgrave Macmillan imprint is published by the registered company Springer Nature Singapore Pte Ltd.
The registered company address is: 152 Beach Road, #21-01/04 Gateway East, Singapore 189721, Singapore

Paper in this product is recyclable.

For the peripatetic sojourners, dedicated social workers, nurturing plant parents and loyal kopitiam patrons of Gaia, whose stories have woven the fabric of this book's existence—Melissa Liow Li Sa

With deep appreciation and love, this book is dedicated to my father, Sam Sui Cham, whose devoted care has been the foundation of our family's strength—Sam Choon-Yin

Preface

In Singapore, neighbourhoods are designed to be walkable, allowing residents to perform daily activities such as grocery shopping, taking their children to school and visiting parks without the need for a car. Places of worship and dining options are situated within proximity, making them easily accessible by foot or public transportation. This is not always the case in other countries, where suburban neighbourhoods may be dominated by houses and require long car trips to access necessities like shopping centres and restaurants. As a result, walking and social interaction are often less common, and many residents may not even know their neighbours.

The concept of walkability has occupied a key role in connecting urban design, architecture and transportation to broader issues like public health, public safety, climate change, sustainability and economic development. The word walkability is associated with *performance or outcomes*, *conditions or means*, otherwise as a substitute of *better urban spaces*. Walking is dependent on the support provided by the built environment. Safety, comfort and scenic landscape matter as well.

Walking can be approached in two ways. The first is to view it as a mode of transportation, used to travel from one place to another, often in conjunction with other modes of transportation such as cars. The second is to view walking as an activity, where the focus is on the act of walking itself rather than reaching a destination, often for leisure or recreational purposes.

There are several types of walking including volksmarching, hill-walking, hiking, trekking, bushwalking, Nordic walking, beach walking and pet walking. Walking has been a longstanding tradition and an environmentally sustainable mode of transportation, particularly in urban areas where space is limited, and congestion is a significant issue. Walking is a zero-emissions mode of transportation that does not require any fuel or infrastructure, making it an affordable and eco-friendly option. By promoting walking as a mode of transportation, cities can reduce their carbon footprint and improve air quality.

Walking is also an affordable and accessible leisure activity that provides positive physical and emotional benefits. Indeed, in a fast-paced environment, many individuals seek solace in walking, whether it's a short stroll or a longer distance trek. Walking offers a chance for people to relax, unwind and relieve mental health issues, while also helping to prevent weight gain and manage life pressure. For some individuals, walking is a recreational activity, providing an opportunity to enjoy the outdoors, while others appreciate its physical and endurance aspects. It is a low-impact physical activity that can be enjoyed by people of all ages and fitness levels.

This book examines how Singapore, a city-state, balances the demand for the use of public space vied by different groups of users as it strives to raise urban walkability. Located at the southern tip of the Malaysian peninsula and just over 100 kilometres north of the equator, Singapore consists of some 63 islets within its territorial waters, constituting a total land area of about 700 square kilometres. With a total population of 5.7 million occupying this small area, Singapore has an extremely high density of 8,300 per square kilometre. A key aspect of Singapore's approach to walkability is its focus on green spaces. Singapore is blessed with abundant green space, which makes it a pleasure to walk around the city. The city has a number of parks and green areas that provide a welcome respite from the hustle and bustle of urban life. These spaces are not only great for recreation but also help to improve air quality and reduce the urban heat island effect. To take an example, on 13 November 2022, the National Parks Board launched the Rifle Range Nature Park. The park is the eighth addition to the central catchment and Bukit Timah Nature Reserve to allow for forest regeneration and serve as home to animals. A key feature of the 66-hectare park is the length of trails, with more than 7 kilometres of trails and boardwalks, the greatest length of trails of all the nature parks in the central catchment area. These parks are designed to be inclusive, with amenities that cater to a wide range of interests and abilities. They are well-maintained and clean, which makes them pleasant places to spend time in. Dedicated walking trails are not just another form of exercise like walking down a city street, which puts demand on attention like navigating through crowds and dodging cars. It puts less demand on attention thereby a more effective way to replenish energy, restore memory and improve cognition.

The idea of transforming Singapore into a Garden City was realised when it was incorporated into the initial Concept Plan in 1971. The Ministry of National Development set up the Garden City Action Committee (GCAC) in 1973 to spearhead the formation of a greening policy, with the Permanent Secretary of the MND as its chairperson. The GCAC prioritised cooling the environment by planting large, tall shade trees such as Rain Trees and Angsana Trees as well as smaller trees like Wild Cinnamon and Hop Trees. The focus was on providing shade to pedestrians along footpaths near the roads, rather than using decorative and colourful shrubs.

To improve Singapore's landscape, the Parks and Recreation Department (PRD) was tasked with new plantings, re-landscaping and high standards of horticultural maintenance. The Public Works Department (PWD) identified locations where trees could be planted along footpaths, while ensuring the continuity of pedestrian walkways. This was achieved by digging up concrete slabs at specific intervals to create planting spaces, which were then covered with concrete slabs to maintain walking paths. Singapore's pedestrian-friendly space and infrastructure are noteworthy. The city-state has developed an extensive network of sidewalks, crosswalks and pedestrian bridges that allow people to move around the city safely and comfortably.

The Housing Development Board (HDB), responsible for constructing public housing in Singapore, also incorporates walkability features into its design guidelines. HDB's town design philosophy extends beyond the physical confines of its flats and extends to the surrounding environment, with a strong emphasis on greenery guided by biophilic principles. Linear green corridors are incorporated for improved connectivity and opportunities for jogging and cycling. Water bodies are designed not only for functional drainage requirements but also to complement the landscape and promote biodiversity, contributing to the recreational features of the town. As part of the city's efforts to become car-lite, HDB collaborates with other government agencies to encourage the use of public transport, walking and cycling. Pedestrian-friendly footpaths and cycling networks connect towns to parks, while park connectors promote walking and cycling. HDB housing estates provide spaces such as sheltered community plazas and community clubs for activities and casual interaction, all within walking distance of the housing blocks and directly connected to public transport modes.

Park Connector Network (PCN) is established by forming regional links and loops (central, eastern, northern, north-eastern riverine, southern ridges and adventure) across the island city that contours a unique character for each of these regions. Residents and tourists alike can easily hop from park to park, enjoying the natural beauty and various activities available. The PCN features beautifully landscaped canals, reservoirs and rivers that connect the major parks and offer some of the best experiences for runners, joggers, brisk walkers, cyclists, nature explorers and rollerblading enthusiasts. It's not only a great way to stay active and connect with nature, but it also provides opportunities for residents to interact with other residents, tourists and businesses, all while enjoying the many attractions that Singapore has to offer.

HDB taps on smart urban solutions and innovative technologies. Smart HDB Town Framework, established in 2014, maps out how information and technology can be used to develop smarter HDB towns, making them more liveable, efficient, safe and sustainable. For example, designers use simulations to study the effectiveness of rain screens in reducing the amount of rainwater that enters a community space. Using wind simulation studies, the alignment of footpaths is optimised to maximise the flow of breezes, providing a pleasant walking experience. Similarly, sun studies are conducted to identify shaded areas for garden trails and habitat, as well as activity zones such as playgrounds and fitness centres, which encourage residents to explore their neighbourhoods. Real-time information on the environment is captured with the installation of sensors. For instance, smart fans in the Neighbourhood Centres of Northshore District can be activated based on data from these sensors related to human traffic, temperature and humidity. This not only enhances the comfort of HDB towns for residents but also helps in reducing energy consumption. Furthermore, mobile apps have been developed to establish a digital ecosystem within the community to generate data and provide new insights for HDB to gain a better understanding of residents' requirements. By using anonymised data on shopping patterns, HDB improves town planning for different types of shops. These efforts are part of HDB's walkability initiatives, which leverage science and technology to enhance urban liveability.

Tengah is the first HDB town in Singapore to be fully integrated with the environment. Tengah boasts a 5-kilometre long and 100 metres wide Forest Corridor, surrounded by a continuous stretch of lush greenery that forms a Forest Fringe, enveloping the town in a serene and natural setting. Anchoring the heart of Tengah is a 20-hectare Central Park, featuring a Forest Amphitheatre and tranquil water bodies that provide a picturesque backdrop for various recreational activities. The Community Farmways, which connect several housing districts within the town, promote walking and cycling and foster social interaction among residents through shared activities such as exercise, farming and gardening. Tengah is Singapore's first car-free town centre, with roads located underground to provide ample space at the ground level for pedestrians and cyclists to travel safely and seamlessly throughout the town. The government has also established "car-lite" zones in certain areas of the city, where pedestrian-friendly amenities take priority over cars. For instance, the Marina Bay area has been transformed into a vibrant pedestrian hub with wide sidewalks, public art installations and green spaces.

Singapore National Parks Board curates islandwide routes across the island with park connectors and trails to offer breath-taking scenery and opportunities for community and tourists to explore outdoor spaces and walk or cycle for short and long distances. Along these routes, we find business entrepreneurs from the foodservices sector to the bike industry offering their services to the visitors. They provide various attractions to enhance the vibrancy and overall enjoyment of the vistior experience. The first phase of the Round Island Route (RIR) spanning 150-kilometre route covering the east coast of Singapore and ending at Berlayer Creek near Labrador Nature Reserve was completed in January 2022, allowing visitors a seamless ride or ramble from the northeast to the south of Singapore. Visitors can make their way to The Oval at Seletar Aerospace Park from Rower's Bay that unravels its rich colonial history and their next point of destination in 3 kilometres that highlights three Heritage Trees and indulging in the gastronomic experiences in the area. Upcoming plans in the pipeline across the Rail, Eastern, Central, Bukit Timah-Rochor Green, C2C Southern and C2C Northern route plans aim to offer a holistic and harmonious urban green space.

To enhance the walkability of a city, cycling is often incorporated through various means. Indeed, the term "active transportation" encompasses all non-motorised modes of travel, including walking and cycling. By adopting a broader perspective that emphasises the benefits of active transportation, planners and policymakers can create more sustainable and liveable communities that prioritise human-powered mobility over car-centric infrastructure. In the case of Singapore, this includes the provision of dedicated cycling lanes and paths that are separated from pedestrian walkways, thereby increasing safety for both pedestrians and cyclists, while also encouraging more people to cycle as a form of transportation. In addition, promoting cycling as a sustainable and healthy option for short-distance travel can help to reduce car usage and traffic congestion in urban areas.

The Walk2Ride programme is a critical initiative under the Land Transport Master Plan 2013 that encourages people to use public transportation by making walking and cycling to and from public transport nodes more convenient and attractive. This initiative involves building covered walkways, installing pedestrian crossings and providing amenities such as benches and bicycle parking facilities. Another scheme called "Green Man + " employs technology to ensure that pedestrians and cyclists have enough time to crossroads safely, especially at busy intersections. The government has also been investing in the creation of park connectors and green spaces, which not only promote walking and cycling but also enhance the city's overall liveability.

In 2015, the Active Mobility Unit was established to oversee walking and cycling policies and initiatives in Singapore. Two years later, LTA passed the Active Mobility Bill to allow the use of bicycles and Personal Mobility Devices (PMDs) on public paths, effectively permitting sharing of footpaths by pedestrians, cyclists and PMD users. More bicycle parking space has been added. This has been achieved by converting car parking spaces into a bicycle parking space, for example, in Telok Ayer Street, Boon Tat Street, Tras Street and Bukit Pasoh Road. Each car parking space converts to 10 bicycle parking spaces. About 15 kilometres of cycling path has been developed in the CBD area, mainly around Bayfront, Marina South and Gardens by the Bay with an additional 3.8 kilometres constructed along Marina Boulevard, Shenton Way, Cecil Street, Cross Street, Pickering Street and parts of South Bridge Road and North Canal Road. The off-road cycling paths are typically between 2 metres and 2.5 metres in width, with coloured pavement (usually red, blue or grey). Intra-town cycling networks are implemented in several townships in Singapore to provide a safe and convenient way for cyclists to commute relatively long distances with the current length of cycling path islandwide exceeding 450 kilometres.

In 2018, a Walking and Cycling Design Guide was published to encourage fair and improved mobility options among the business community, residents and tourists. The guide maps a clear framework and defines the necessary infrastructure to enable walking, cycling and riding on personal mobility devices (PMDs) from the first to last kilometre. The Singapore government advocates a healthier lifestyle in conjunction with Singapore Healthy 365 mobile app roll-out that includes a fitness tracker and incentives redeemed through Your Healthpoints (HP) for TapForMore (TFM) points or TransitLink eVouchers. The guide offers a common set of design guidelines for consultants, private developers, the building industry and government agencies by enabling active mobility and usage of associated infrastructure in a harmonious way.

As evident, Singapore places a significant emphasis on walkability in urban planning, and several policies and initiatives have been put in place to achieve this goal, resulting in a well-maintained environment that is both clean and green. Singapore has invested heavily in creating a pedestrian-friendly environment that caters to people of all ages and abilities, creating a pedestrian-friendly environment, with an extensive network of footpaths, wide sidewalks, shade structures and pedestrian crossings at regular intervals. This makes it easier for people with disabilities or mobility issues to get around, as well as for families with young children and elderly residents. While construction works may present obstacles to pedestrians from time to time, many of the pavements are designed to be wheelchair friendly, ensuring that they are accessible to people of all abilities. Singapore's extensive public transportation network, which includes buses, trains and light rail, makes it easy and affordable for people to get around the city without relying on cars or other forms of private transportation. The emphasis on urban planning that prioritises walkability can have several positive outcomes, including improved public health, reduced traffic congestion and enhanced economic activity. Singapore's focus on walkability has helped to create a vibrant and dynamic city that is both sustainable and welcoming. As other cities around the world look to improve their urban environments, they would do well to look to Singapore for inspiration.

Singapore, Singapore

Melissa Liow Li Sa
Sam Choon-Yin

Acknowledgments

Collaborating on the creation of this book has been an extraordinary and enriching journey. Each of us has contributed our unique perspectives and insights, intertwining them into a harmonious and meaningful narrative. As former colleagues and friends, our shared passion, and resolute commitment to a common vision have seamlessly merged to give life to the pages of this book. This collaborative effort has allowed us to not only put words on paper but to also forge connections and insights that transcend the pages. We are profoundly grateful for the opportunity to bring this project to life, a true embodiment of our dynamic partnership.

The creation of this book owes its existence to the unswerving support of numerous individuals. We extend our heartfelt gratitude to scholars who have played an instrumental role in moulding this work into its current form: Dr. Eunice Tan Mei Li and Dr. Pauline Anne Gordon. Their steadfast commitment, insightful perspectives and constructive critiques have elevated this work to a remarkable level. We are profoundly appreciative of their time, dedication and invaluable contributions. This book stands as a testament to our shared devotion and pursuit of excellence.

Our sincere gratitude extends to the Palgrave Macmillan team. We would like to thank Dr. Hua Bai for recognising the potential of this project, Ananda Kumar Mariappan for expertly guiding us throughout the journey and Nandakini Lahiri for enhancing our ideas and writing style. We highly value their indispensable recommendations and continuing support throughout the entire production process. Their contributions have been truly invaluable to us.

Melissa would like to extend her sincere appreciation to her colleagues, specifically mentioning Derrick Chang, Vincent Chong, Pearly Wong, Dr. Charles Ong, Dr. Calvin Tan and Cecilia Yeoh. Their outstanding leadership has served as a guiding light, illuminating her path through the intricacies of the professional domain. Their invaluable insights, mentorship and belief in her abilities have given her the courage to tackle challenges and pursue excellence with firm determination.

This book conveys that the viewpoints presented in its pages are the sole expressions of the authors. Our words and concepts form the foundation of the narrative, offering readers a direct insight into our distinct viewpoints, beliefs, and insights. Consequently, it mirrors our individual intellectual voyage and thought progression. Readers are urged to approach the content with an open-minded attitude, valuing the varied array of perspectives that enrich the discourse on matters pertaining to sustainable urban development.

Last but not least, we extend our heartfelt gratitude to our families whose unwavering support and encouragement have been our pillars of strength throughout the book-writing journey. Melissa would like to express her deep appreciation to her beloved parents, Liow Sun Lai and Kung Kiat Aim, who now reside in heaven. This book is a tribute to their enduring love, guidance and the values they instilled in her. Special recognition goes to her husband, Dr. Yeow Kim Chai, her brother, Irwin Liow Li Ming and son, Yeow Chung Han, for their unshakeable belief in her and their constant presence, which have enriched her life incredibly.

Sam is immensely thankful to his parents, Sam Sui Cham and Choo Lai Har, for their constant love, care and the educational opportunities they provided. He extends heartfelt thanks to his wife, Shiau Hong, and his children, Mun Wing and Wai Lok, for their unending love and unwavering support throughout his writing and reading pursuits. Their encouragement has been a driving force behind his efforts, and their presence has brought immeasurable joy to his life.

August 2023

Melissa Liow Li Sa
Sam Choon-Yin

About This Book

The structure of the book is as follows.

Chapter 1 explores the legacy of walking as a mode of transportation. While walking and cycling have been more commonly embraced in European and North American cities, the tide is shifting in some Asian cities. The growing awareness of the environmental and health benefits associated with walking is inspiring a change in mindset across Asia. In line with the principles of sustainability, cosmopolitan cities like Singapore are actively promoting public transportation, reducing car dependency and fostering pedestrian-friendly urban environments.

Delineating public space for walking in Singapore's concrete jungle and state-community power relationships in the colonial, post-colonial and post-post-colonial periods is discussed in Chapter 2. Land supply in Singapore is the biggest constraint and therefore must be put to the best use. When Singapore gained self-governing status in 1959, there was severe overcrowding and housing shortages. Developing Singapore into an economic hub to compete at the regional and international levels was a top priority. Cycling paths were removed to widen the roads to serve the needs of motorists. The city focused on developing good infrastructure to offer sufficient capacity and interconnectivity for transport, logistics and supply chains.

Chapter 3 highlights the government's attempt to promote walking and cycling as the main modes of active mobility in *contemporary* Singapore to offer Singapore a way to deal with the urban transportation challenges and simultaneously promoting sustainability, health and quality of life in the city-state. In land scarce Singapore, reducing the demand for cars makes sense, considering that each car occupies two to six parking spaces at home, office and other destinations, which takes much more land comparatively speaking than walking and cycling. Chapter 3 also explores cycling as an alternative mode of transport. Singapore aims to be a "car-lite" travel city, thus implements tough efforts to entice residents to cycle. To encourage more people towards a transport alternative, it is imperative to comprehend their attitudes, interests and opinions towards active mobility and formulate appropriate transport network and schemes. Bike-sharing schemes, end-of-the-trip facilities, public transport and feeder service connectivity, flexible working hours and education for all road users to enable sharing and high safety assurance, are some measures that attract people to cycle in Singapore.

Chapter 4 focuses on the heritage buildings and monuments in Singapore that represent the nation's culture and identity. Singapore is an interesting case to show what can be accomplished by a small country with scarce natural resources. A discovery of some of Singapore's historic ethnic enclaves and themed walking trials in its Garden City are some attractions tourists can bask themselves in this state as we show in Chapter 4.

Chapter 5 focuses on food tourism as an integral part of Singaporean culture. Through the lens of gastrodiplomacy, the chapter highlights the famous hawker culture, which has been recognised with the UNESCO's List of the Intangible Cultural Heritage of Humanity title. Singapore offers hawker food in both traditional and modern outlets, with some having received one-to-three Michelin star ratings. The city hosts a range of food-related events, such as cooking events and gourmet festivals, in its famous historic ethnic enclaves. The chapter explains that Singapore's food culture plays a significant role in all three dimensions of the brand resonance model: awareness of Hawker Culture, love for Hawker Culture and loyalty to Hawker Culture. Food tourism can be connected to walkability in several ways. When a city has a vibrant food scene, it can encourage visitors to explore different neighbourhoods on foot, which can contribute to the walkability of the city. Walkability can also be enhanced by the presence of pedestrian-friendly areas and infrastructure such as sidewalks, street crossings and well-lit pathways, which can make it easier and more enjoyable for visitors to walk to and from various food destinations.

Chapter 6 burrows into Singapore's parks, trails and gardens rich with natural heritage and biodiversity, despite better known to the world as a highly urbanised metropolis. Fuelled by prolific human desire to reconnect with nature as people are increasingly living in isolation from nature, there are the emerging lifestyle patterns of a diverse range, including walkability, hiking, trekking and cycling. Effective interpretation is crucial in forming meaningful human–wildlife interactions, while concurrently conserving wildlife habitats.

The chapter also covers Singapore's spatial reorganisation of its nature reserves and islands to meet the needs of its urban core. Initially, this was a difficult undertaking, as the nation had reclaimed land from the sea to expand its coastlines and cleared some forest areas for urban development. Communities were displaced, and reef colonies destroyed. However, despite these challenges, many activists, scientists and civil society organisations are working to conserve the remaining natural and cultural heritage of the reserves and islands. Additionally, new forms of rustic tourism and leisure have emerged as Singapore transitions to becoming a "City in Nature".

Chapter 7 focuses on the role of managing vehicle population in promoting sustainable development in Singapore. The nation encourages its citizens to adopt a car-lite lifestyle, using public transportation, walking and cycling as primary modes of transportation. This strategy reduces traffic congestion and frees up land space, contributing to a more inclusive, safe, resilient and sustainable environment. Presently, the Singapore government regulates the number of cars on the road through the release of Certificate of Entitlements (COEs). The country's well-developed public transport system ensures that its people have access to affordable, reliable, sustainable and modern energy. The government has invested heavily in developing clean energy ride-hailing services and public bus operations. The Mass Rapid Transit (MRT) and Light Rail Transit (LRT) systems provide an affordable and convenient alternative to car ownership and are the backbone of Singapore's public transportation system. In line with the Green Transit-Development (TOD) concept, the construction process of the MRT and LRT considers sustainable development, promoting the creation of vibrant, walkable neighbourhoods with access to various amenities and services, such as parks, shops and schools. TOD developments typically include a mix of residential, commercial and office spaces, with the aim of reducing commuting times and promoting sustainable modes of transport such as walking, cycling and public transportation.

That a robust economy is needed to ensure Singapore's survival as a nation is a well-accepted premise. In this regard, urban planning and economic planning are related and connected. Tree planting exercise, an initiative started in November 1971, and nature conservation efforts could be described as economically motivated. As stated by the first Prime Minister, "in wooing investors, even trees matter". A clean and green Singapore helps to enhance economic performance through the arrival of tourists, "green" companies and foreign investors attracted to a well-organised place, although the initiative also helps to conserve the environment by removing carbon dioxide from the atmosphere. Economic planning helps to shape the Singapore economy. Chapter 8 examines the role of urban design to facilitate economics activities and movement of people to support economic developmental goals.

Besides the physical environment, Singapore uses integrated Information and Communication Technologies (ICTs) and big data on cyclist and pedestrian behaviours to improve active mobility. Walkability enhanced ICT provides routing, locational data and real-time information to facilitate walking and cycling thereby forms a crucial part of walkability. ICT, if employed effectively, can exert influence on walking and cycling activities. Chapter 9 outlines the rapid development of ICTs, catering to pedestrian and cyclist needs and preferences with reference to Singapore's effort to develop ICTs vigorously as part of its effort to establish itself as a smart city.

Chapter 10 focuses on tourist mobility, which demonstrates positive gains. Tourists *escape* from their normal hectic lives and explore the unconventional *off beaten* tracks or travel to places that are under-explored. By motivating people to take-up physical activities, negative impact of motorisation can be mitigated. Underpinned by the mobility concept, the mobile daily life activities weave unique leisure and tourism experiences as well as institute community engagement. To some extent, cycling tourism is similar to walking, enabling a close connection with nature and interaction with the local population. This chapter seeks to understand tourist mobility in Singapore and how the active mobility network has become a part of Singapore's agenda in promoting and growing the tourism sector in Singapore.

Chapter 11 surfaces the matters surrounding population ageing which has become a major public policy concern globally, and Singapore is no exception. One in four Singaporeans will be 65 years old and above by 2030. New realities resulted from the demographic shift—changing family structures with more seniors staying alone, a shrinking workforce and increased healthcare needs—would be evident. If Singapore plans well for these changes and enables seniors to lead healthier, happier and more productive lives, the nation can gain the silver dividends that come along with increased longevity. An Urban Systems Study marks Singapore's journey in planning for an ageing population, documenting the initiatives implemented from the 1980s. To ensure that planning continues to strive in satisfying seniors' needs and interests, the government boarded a year-long consultation process in 2014 comprising some 4,000 residents from different walks of life, to gauge their aspirations for ageing. This ultimately climaxed into the development of a national blueprint that encompasses the Action Plan for Successful Ageing including public spaces, transport, employment, health and wellness to make the country a City for All Ages, and more so a better place to age.

Chapter 12 discusses Singapore's planning methodologies, from a modernist approach to embracing biophilic strategies. To accomplish the vision of the Garden City with sustainable development in mind, the nation is projecting an image that focuses on social needs and quality environment. This requires a multi pronged approach, which includes aligning the vision with key agencies to allow for coherent efforts and targeted outcomes—to develop the green infrastructure within the urban spaces. Singapore has committed to three horizons of urban sustainability performance—green design, sustainable development and regenerative urbanism. The aim is to progress to a state of operating urban infrastructure and systems that are kind to the environment, and at the same time, making walking and cycling amid its congested urban life pleasant and environmentally viable.

The development of a city is a complex process. It involves balancing interest, strong leadership and engagement with the stakeholders, building frameworks and institutional capacity to ensure plans and projects are carried out effectively. Detailed plans have to be designed and implemented cross institutionally possibly over different terms of the government due to the long gestation period. Are there lessons to be drawn from Singapore's experience? Instead of copying the model wholesale, it is possibly more useful to select relevant aspects of the model and adjust them to suit their local environment. Chapter 13 adopts this approach to highlight a number of lessons which urban planners around the world could *emulate* to envisage walkability in an urban concrete jungle.

Contents

About the Authors

Melissa Liow Li Sa is currently the Assistant Head of the School of Business and Management at PSB Academy. She graduated with a Bachelor of Business Degree from the University of Technology Sydney (UTS), Masters in Business Administration from the University of Wales (UOW) and Ph.D. in Business from the International University of Malaya-Wales (IUMW). She has taught subjects in Marketing, Human Resource Management and Tourism and Hospitality Management. Her research interests lie in the field of tourism entrepreneurship. Aside from her responsibilities in teaching and research, she contributes as the Programme Leader for the Hospitality and Tourism programmes at PSB Academy. She also contributes as Ex Officio of Assessment Committee at PSB Academy.

Sam Choon-Yin is currently the Dean of Academic Partnerships at Kaplan Higher Education Institute, Singapore. He graduated from the National University of Singapore (NUS), University of Technology, Sydney (UTS), and University of South Australia (UNISA) where he obtained his Ph.D. in International Business and Management. His research papers have appeared in journals such as Asian Studies Review, Journal of the Asia Pacific Economy, South East Asia Research, Global Economic Review, New Zealand Journal of Asian Studies, Journal of Asian and African Studies, and Asia Pacific Viewpoint.

Abbreviations

EDB	Economic Development Board
HDB	Housing Development Board
IDA	InfoComm Development Authority of Singapore
IMDA	Infocomm Media Development Authority
JTC	Jurong Town Corporation
LRT	Light Rail Transit
LTA	Land Transport Authority
MDA	Media Development Authority
MRT	Mass Rapid Transit
NCB	National Computer Board
NEA	National Environment Agency
NParks	National Parks Board
PAP	People's Action Party
PUB	Public Utilities Board
STB	Singapore Tourism Board
URA	Urban Redevelopment Authority

List of Tables

CHAPTER 1

Walking Traditions Around the World

The concept of walkability has captured the attention of urban planners in conjunction with efforts to make cities sustainable, pedestrian friendly and integrated with the public transportation system. It is understood that walkability is dependent on a host of factors. Proximity of public transport stops and key locations in the city, and efficient public transport system (comfortable and high frequency) are more likely to increase the use of public transport and therefore the opportunity for walking to take place. Although significant increase in motorised traffic due to urbanisation and economic growth has disrupted the longstanding practice of walking in cities, high traffic zone with cars overflowing on the street and lack of designated parking slots also translate to greater convenience in taking public transport and making walking more attractive. There is much effort to promote walking and cycling in Asia and elsewhere, including measures to curtail the dominance of car vehicles and increase pedestrian friendliness in cosmopolitan cities.

But first, who is a pedestrian? Pedestrian, according to the Sixth Edition of the Little Oxford Dictionary, can be referred to as "a walker, traveller on foot", in which case pedestrian activity is a form of transportation comparable to other modes of transport like driving, cycling and rail and sea transportation. The American Heritage Dictionary expands the definition of those walking to include non-transportation purposes

Melissa Liow Li Sa and Sam Choon-Yin, *Sustainable Urban Development in Singapore*, Sustainable Development Goals Series,
https://doi.org/10.1007/978-981-99-5451-3_1

such as exercise, recreation, leisure, shopping, social interaction, spiritual rejuvenation or even fundraising for charity.[1]

What is walking? According to Giovanni Cavagna, Norman Heglund and Richard Taylor, walking or ambulation among homo sapiens are gaits that require balance and muscles coordination so that the body is thrusted in an inverted pendulum stride with each step.[2] The word *walk* is descended from *wealcan* or *to roll*, which can be traced back to the Anglo-Saxon days in the middle of the fifth century.[3] Walking is different from running as there are times of double support when standing and that only one foot at one time is lifted from the ground with a step. Speed is the other aspect that characterises walking from running. An infant gradually gains independent walking ability at about 11 months old or so.[4] Chang Chien-Yen and his colleagues concluded that for a single child, the mean walking speed is 1.15–1.17 metres per second and the 15th percentile walking speed is 0.98 metres per second.[5] The mean walking speed for a single elderly is comparatively slower than a single child, at 0.89 metres per second and at 0.73 metres per second for the 15th percentile walking speed.

There are several types of walking—volksmarching, hillwalking, hiking, trekking, bushwalking, Nordic walking, beach walking and pet walking. Australians bushwalk, Canadian's hike and people in South Africa as well as in the United States tend to go for long active walks while in parts of North America, one is more indulged in short walks around towns and cities. It is just walking or hillwalking in UK and Australia, but similar walks are known as tramps in New Zealand. Some people use a pedometre to count steps while others might prefer to walk indoors on a treadmill at home or in a gymnasium.

Rambling is a walking activity that ranges from leisurely short strolls to more challenging mountain treks, typically taking place in the countryside. The largest walkers' entity in Britain is the Ramblers' Association, with a history since 1935 and has over 400 local groups and 130 thousand members in Great Britain. Another association is the Long-Distance

[1] Quoted in Lo (2009, pp. 145–146).

[2] Cavagna et al. (1977).

[3] Biewener and Patek (2003).

[4] Samra and Specker (2007).

[5] Chang et al. (2010).

Walkers Association (LDWA), a British not-for-profit, volunteer-led association founded in 1972. The entity has close to 10,000 members dedicated to walk a distance of between 20 and 100 miles annually. These walks are performed along long-distance paths over one to several days. Some of the world's well-known rambling trails include the four-day, 45 kilometres Inca Trail, Peru (Latin America and the Caribbean); five- to seven-days 75 kilometres West Coast Trail, Canada (North America); three- to five-days 45 kilometres Dana to Petra, southern Jordan (Middle East); six-day 60 kilometres Tsitsikamma Trail, Eastern Cape of South Africa (African continent); four-day, 53.5 kilometres Milford Track, New Zealand (Australasia); 150-days, 1,700 kilometres Great Himalaya Trail, Nepal (Asia) and nine to twelve-day, 170 kilometres Tour du Mont Blanc, France, Switzerland and Italy (Europe).

The scientific evidence is unequivocal in demonstrating the significant advantages of walking. Walking helps to maintain an individual psychological well-being in the form of rejuvenation to beat mental fatigue and stress and lower the risk of diabetes, cancer and cardiovascular disease. Sarah Hanson and Andy Jones found that walking groups demonstrate significant increase in SF-36 (physical functioning), VO2max (oxygen utilisation during exercise) and a six-minute walk time of 79.6 m (53.37–105.84) and significant reduction in depression, resting heart rate, systolic and diastolic blood pressure, total cholesterol, body mass index and body fat.[6] Gunnthora Olafsdottir, Paul Cloke and Andre Schulz examined the impact of walking in nature compared to watching nature scene on television. The former demonstrates restorative influences on cortisol levels and helps improve people's moods.[7]

Walking is particularly useful in busy and fast-paced lifestyles, which are progressively portrayed by deskbound behaviour, life pressure, weight gain, mental health issues and detachment from nature. In a knowledge-based economy such as Singapore where many people feel that their lives are speeding up, walking and cycling offer ways to balance the flow of everyday life, one that is less hurried.[8] The creative class, for example, seeks to recharge through physical activity. Richard Florida wrote, "If you

[6] Hanson and Jones (2015).

[7] Olafsdottir et al. (2020).

[8] Walking speed is probably a good indicator of the pace of life. An interesting international study from the 1990s showed as the general pace of life sped up in cities, city dwellers were more likely to walk faster (see Levine, 1997).

spread your workday in front of a computer screen or an artist's canvas, you probably are not eager to spend your leisure in front of a TV screen. You are much more likely to want to get out and be active", peddling a bike or walking.[9]

It is worth noting that some people may choose to walk and take public transport not because they want to, but mostly due to their inability to own a private car or take a cab. This travel characteristic particularly holds true for the lower socioeconomic group. In this regard, walkability among car owners is of vital importance—to switch to less energy-intensive but more time-consuming mode of transport—if walking is to increase significantly as a requirement will be to lessen the impact of deterring factors and alter the negative perception of walking. To build an environment to support and encourage walking, urban planners have focused on providing pedestrian comfort and safety, connectivity with destinations and richly visible landscapes. Public spaces flanked by buildings offer hospitable pedestrian path with street trees, bus shelters, street art and path lighting to enhance night-time walking and provide greater security. To encourage people to explore the city streets on foot, vacant spaces locate parks with small retail shops. In New York's High Line, people walk through gardens, savour delicious foods, enjoy a performance and view art in the company of friends and neighbours. Atlanta boosts a more recent Beltline and probably the largest greenway in America, sprawling with eight parks, 7-mile trails and transit which links to 45 neighbourhoods.

Notably, cities are going car-lite, in the move to support walking and other public transportation modes as well as to reduce their carbon footprint. Efforts to develop *pedestrian villages* as part of urban development efforts were targeted to make the places pedestrian friendly with cars parked below in underground basements or the boundary of the village. Going car-lite is not a new thing for Singapore. The Sustainable Singapore Blueprint unveiled in 2015, detailed a 15-year plan to reduce the usage of cars to free up space for shared community spaces, pedestrian and cycling

[9] Florida (2002, p. 182).

paths, parks and greenery.[10] ELTIS, a European portal for local transport, also shares a range of good practices around the world, detailing how walking has gained traction as a mode of transportation in cities and towns. In particular, infrastructure such as river walks, foreshore ways, ocean ways, cafes, boutique shops and malls play a vital role in increasing participation and enjoyment of walking. Some of the famous pedestrianised streets are Strøget in Copenhagen, Denmark, which became a car-free city back in 1962; Rue Mouffetard in Paris, France, dating back to the twelfth century when the city served as the Roman road. Others include Carnaby Street in London, England, which was pedestrianised in 1973; Qianmen Street in Beijing, China, in 1965 (which houses Chinese architecture and the old history tracing back to the Qing and Ming dynasties) and Flower Street in 1972, a pedestrian-only boulevard in Curitiba, Brazil.[11]

Most politicians and urban planners know very well that the success of the city is powered by its ability to attract tourists from around the world. In the travel and tourism world, the opportunities for economic gains encompass organised trekking holidays (for example, the Annapurna Circuit within the mountainous regions of central Nepal) and guided sightseeing walk tours in cities and towns. Well-organised trail systems are thriving, spanning from Australia and New Zealand to Canada and the United States. Waymarked walking trails across the European continent are also found in nations as far as Cyprus, Norway, Portugal and Turkey. Many also undertake the epic pilgrimage walks, of which the most well-known is the Kumono Kodo in Japan, the Abraham Path in the Middle East, the Pilgrims' Way in England, El Camino de Santiago

[10] In November 2018, LTA announced the new Range-based Parking Provision Standards (RPPS) and the new parking Zone 4 for car-lite areas, which came into force in February 2019. The number of parking spaces in these areas has been significantly reduced to free up space for shared community spaces, pedestrian and cycling paths, parks, and greenery. Classified as "Zone 4" in the RPPS, the car-lite areas gazetted on 1 February 2019 were Kampong Bugis, Marina South, Jurong Lake District (JLD), Bayshore and Woodlands North. On 1 August 2020, additional five new car-lite areas were gazetted—Jurong Innovation District (JID), one-north, Punggol Digital District (PDD), Springleaf, and Woodlands Central. On 5 October 2022, five additional areas were gazetted, as car-lite. Among these, four include public housing estates—Ulu Pandan, Mount Pleasant, Tengah and Keppel Club site (which forms part of the Greater Southern Waterfront).

[11] Cappello and Buder (2020).

in Spain as well as La Via Francigena in England, France and Italy. Hallmarked walking events and festivals are organised actively in countries such as the Vierdaagse International Four Day Marches in the Netherlands since 1916, International Gilboa 2-day Walk in Israel since 1999, Canberra Walking Festival in Australia, as well as Hærvejsmarchen in Virbog, Denmark. Walking buddies in these walking events and festivals march as far as 30 to 50 kilometres per day for two to four days straight.

To form a basis for comparing different cities around the world with regard to walkability, Walk Score develops the walkability index (*Walk Score*) on a scale of 0 to 100 to measure and rank the walkability of cities and locations. A higher score denotes greater walkability. According to the Walk Score (2021), the top five walkable American cities in 2020 are New York (walk score 88.3), San Francisco (walk score 87.4), Boston (walk score 82), Philadelphia (walk score 79) and Miami (walk score 78). The walk scores recorded in American cities in comparison to Canadian cities are higher with Vancouver's walk score of 79.8, Montreal (walk score 65.4), Toronto (walk score 61.0), Hamilton (walk score 49.6) and Mississauga (walk score 48.9). For the Australian cities, it is established the top five walkable cities are Sydney (walk score 62.6), Melbourne (walk score 57.0), Adelaide (walk score 53.6), Brisbane (walk score 51.2) and Perth (walk score 50.2). World Bank's Global Walkability Index (GWI) includes variables such as crossing safety, perceived security and accessibility. GWI forces on aspects of walkability "that can be improved upon in the short and medium terms" thereby omitting land-use variables and their effects on convenience and connectedness of the pedestrian network (see Table 1.1).[12]

While walking and walkability are highly regarded as beneficial, scarcity of resources forces urban planners to think about the value of one option over another. Enlarging the common space could correspondingly restrict individual space. For instance, if there are additional lands dedicated to pedestrians, roads may be narrowed for motorists. Urban planners are often caught in a dilemma and in some cases resign back into redesigning outdates and unsustainable urban plans. Devesh Doobay, Director for Planning and Economic Development at the City of Riverdale from 2004 to 2019, said:

[12] Quoted in Krambeck and Shah (2006, p. 5).

Table 1.1 Components of the Global Walkability Index (GWI)

Component	*Variable*
Safety and security	Proportion of road accidents that resulted in pedestrian fatalities Walking path modal conflict Crossing safety Perception of security from crime Quality of motorist behaviour
Convenience and attractiveness	Maintenance and cleanliness of walking paths Existence and quality of facilities for blind and disabled persons Amenities, e.g., coverage, benches, public toilets Permanent and temporary obstacles on walking paths Availability of crossings along major roads
Policy support	Funding of resources devoted to pedestrian planning Presence of relevant urban design guidelines Existence and enforcement of relevant pedestrian safety laws and regulations Degree of public outreach for pedestrian and driving safety and etiquette

Source Krambeck and Shah (2006, p. 10)

> Nowadays with real estate, listings include walkability scores.... People want to walk their kids to school, or to friends' houses. People want to walk to the town centre. Even people not in cities want this. Part of the job of urban planners now is to fix things. People used to want the sprawl, which we're used to seeing in most towns and newer cities. But now, we have to retrofit.[13]

In regard to urban design for tourism purposes, urban planners are still not clear on what defines the legit space entitlement to tourists and how they can be motivated to use public transport and walk. Achieving seamless connectivity between hotels, restaurants, shopping strips and places of attractions requires an integrated framework for land use and transport development which continues to evolve as states strive to balance social, economic and environmental interests.

[13] Quoted in Caldwell (2019).

Walking in Asian Cities

Asian cities are conventionally very walkable. Walking is pervasive particularly in the developing countries as it is a cheaper mobility mode. In spite of the prevalent practice and tradition of walking in Asian cities, transportation has become motorised due to urbanisation and economic growth. According to the Asian Development Bank, countries in the Asia Pacific region face dire environmental challenges including air pollution resulting from rapid industrialisation and exploitation of natural resources.[14] The unpleasant environment is worsened by inadequate city planning and poor municipal governance, making the city streets unwelcoming to walkers. Asian cities are sparing more effort to convert more sustainable, healthy and walkable streets and spaces. UN Environment Programme (UNEP) noted that 50% of Asian cities are keen to promote walking at the city and national level, and on average 83% of the cities surveyed had non-motorised transport pledges. Some Asian cities are going car-lite and encouraging greater use of public transportation to enhance the visibility of and accessibility to street-level events. Additional measures are undertaken to widen pavements, better street pedestrianisation and linkages to nice appealing social spaces like Scramblewalk in Shibuya district in Tokyo and the restoration works of the Cheonggyechon Stream in Seoul. High-density urban cities, good public transit, mixed-use environment and strong civic have enabled active mobility and city dwellers to enjoy the benefits of walking and cycling.

According to a review on urban mobility by Centre for Liveable Cities and Urban Land Institute, 8 out of 10 role model cities are in the Asia Pacific region, including Singapore, Bandung in Indonesia, Shanghai, Taipei, Suwon and Seoul in South Korea and Tokyo, as well as Yangon, Ho Chi Minh and Sydney.[15] Travel bloggers have voted their own fair share of their favourite walkable cities in Asia spanning from Malaysia to Mongolia, and Japan to Uzbekistan. Exploring a city by foot in Asia offers the opportunity to bask in a city's local side amid thriving street culture, which is often the case especially in some of the largest Asian metropolises. Some of the highlights include street food, street art, pedestrian zones in old towns and shopping malls. Street trading for urban-based livelihoods and food security is a common sight. Street trading does not occur

[14] Asian Development Bank (2011).

[15] Centre for Liveable Cities and Urban Land Institute (2017).

in neutral spaces, but rather in contested and lived-in spaces, leading to pavement congestion and forcing walkers onto roads.

Street-level activities were once prominent in Singapore. There was a carnival-like atmosphere daily along the streets and alleys, filling with the sound of languages and dialects. Low-price hawkers and eating places operated along the Singapore River were often frequented by lighter men and manual workers. Pollution was a severe problem attributed mainly to the dumping of rubbish by businesses and individuals, including hawkers. The river was treated as a convenient sewer. In 1969, the PAP led by Lee Kuan Yew intervened to identify the sources of pollution and measures to deal with the situation were presented. Many government departments were involved under the scrutiny of the Prime Minister himself. Cleaning up the streets in general and the waterways in particular became a national goal of the new government, culminating in the famous "Clean Up the River" speech by Lee at the opening of the Upper Pierce Reservoir in 1977. Hawkers and others were reallocated and housed in newly constructed hawker centres. Slums and squatters were removed. Residents were allocated public housing units built by the Housing Development Board.

While Asia is stepping up to improve walkability and pedestrian facilities, warm weather all-year round in some parts of Asia, including Singapore and rising temperature during the summer seasons is the primary source of discomfort when engaging in the walking activity. In this regard, street-level intervention is important to make walking comfortable. The green walking corridors and shelter with covered walkways at pedestrian crossings, linking residential places and public transport hubs enable shorter walking distance under the hot sun. Extensive network of green walking paths and park connector networks are built in cities to increase the likelihood of walking. Urban planners are also keeping walking safe and comfortable with properly lit walkways, clear and user-friendly pedestrian walks and installation of modal filters and traffic calming steps which aim to reduce the speed of motorised vehicles (to 20 to 30 kilometres per hour) in high pedestrian traffic areas. In Malacca and Kuala Lumpur, Malaysia, alleyways are spruced up with furniture and art to encourage community engagement. Urban green space programme in Malang, Indonesia, receives funding from public and private corporations to construct or improve greenways and street parklets. Walking environments are built to tailor the needs of stakeholders, henceforth fostering support, ownership and active participation. Asian streets are in large measure communal spaces where people spend time for play and organise community gatherings and informal business meetings. Walking with the use of attractive off-street walking networks with infrastructure against the backdrop of green corridors and waterfronts, with indoor walking routes like the walking pavements along the central parts of the Malacca River is one example.

Walkability and Sustainability

We are living in extraordinary times. Consumption expenditures are rising. In the United States, the fraction of personal consumption was at 61.5% in 1969, rising to 65.6% 20 years later and almost 70% today. More goods move faster across borders resulting from advances in communication and transportation. Such consumption pattern places a tremendous burden on the planet, giving rise to problems such as global climate change and worldwide loss of biodiversity. CO_2 concentration in the atmosphere causes warming of the surface of the earth and melting of permafrost. Drought conditions are spreading to Australia, the United States and others. Climate destabilisation also causes famines and storms, causing immediate loss and disruption, including billions of lost agricultural production.

Indeed, Arctic Sea ice is melting at unimaginable rates. Sea levels are rising. Higher sea level will wipe out small island nations, forcing dwellers to migrate inland.[16] Without a hinterland, Singapore, a low-lying island city-state, is highly vulnerable to the impacts of sea level rise. It is projected that sea levels in Singapore could rise to five metres by 2100, posing a significant threat to the livelihoods and safety of the people. To address this challenge, the Singapore government has taken proactive steps such as launching the Coastal Protection and Flood Management programme in March 2023. This programme aims to support the development of innovative and effective coastal protection and flood management solutions, including the installation of coastal barriers and

[16] The Washing Post reported in January 2021 that Earth is losing 1.2 trillion tonnes of ice each year, citing a report by the American Association for the Advancement of Science (AAAS). See Mooney and Freedman (2021, January 25). Nouriel Roubini (2022, p. 221) noted that China has the largest population residing in coastal areas with sea rise likely to affect 130 million of its citizens. Fifty-five million people in India could see their homes submerged followed by 41 million in Bangladesh who lives near the coast.

the enhancement of drainage infrastructure. By taking an anticipatory and adaptive approach, Singapore is taking critical steps to safeguard the well-being and future of its people, while also demonstrating global leadership in climate resilience.

Cities are responsible for a substantial proportion of global warming. The increasing concentration of human populations in urban areas has resulted in the most significant environmental changes. Despite taking up only a small fraction of the earth's surface, cities are responsible for the majority of the impacts of climate change. The high concentration of people in urban areas leads to high demand for resources, resulting in high carbon emissions. The largest cities in the world, such as Tokyo, Shanghai, Sao Paolo, Mexico City, Lagos, Rio de Janeiro, Dhaka, Istanbul, Moscow, Manila, Mumbai and Jakarta, have more than twenty million people, generating vast amounts of waste and posing significant challenges to sanitation treatment. The consumption and production of food, water and energy are much higher in cities and urban areas, with most of these resources transported from other locations, resulting in excessive energy use and carbon emissions. The factories and manufacturing plants located in or near cities rely on materials from other locations, which must be transported, contributing to global energy-related CO_2 emissions. In some cases, the inefficient production and transportation of goods can lead to even higher emissions.

The United Nations Human Settlements Programme (UN-Habitat) was set up to promote socially and environmentally sustainable towns and cities. An important contribution of the UN-Habitat was to shift the focus of planners to accessibility—destinations, activities, services and goods—rather than adding the quantity and length of transport infrastructure. World Health Organization (WHO) recognises the importance of urban design and transport to improve urban health standards in this and future generations through change to the urban environment. With an estimated 70% of the people likely to live in urban cities by 2050, a well-planned and managed urban development and transportation are essential to create sustainable cities and communities.

Promoting walkability in cities and urban areas is a crucial factor in promoting sustainable and healthy urban development. Walking is considered a green type or sustainable form of transportation, appropriate even for urban lifestyles and for short distances. By providing safe, convenient and comfortable walking environments, cities can encourage residents to adopt active modes of transportation, reduce reliance on motorised vehicles and subsequently decrease traffic congestion, air pollution and greenhouse gas emissions. Urban areas are particularly important to focus on as they are the centres of economic and social activities, and the majority of the global population lives in cities. The high population density in urban areas necessitates efficient and sustainable transportation systems, and walkability can play a crucial role in achieving these goals. The urban environment presents unique challenges to walking, such as safety concerns, the presence of high-speed traffic and a lack of pedestrian infrastructure. Addressing these challenges through urban planning and design is important to enhance the walkability of cities and contribute to the creation of healthier and more sustainable urban environments.

As direct participants in the ecosystems, we must do our part. We much increase our awareness of the environmental problem. Interdisciplinary work in psychology, biology, economics and sociology shows that human beings are adaptable and malleable. Neuroscience tells us that human brains are plastic. We can change how we live our life for the betterment of society. Together, we must find a new route for living, a route that respects and cares for the planet.

The good news is that initiatives for healthier urban development are sprawling around the world. Cities around the world have spent millions developing and creating cultural districts and vibrant street life busted with active outdoor recreation to offer desirable places to live in. Central Park is known as one of the most known green spaces in America and is now permanently close to cars. Los Angeles is undergoing an infrastructure revitalisation to become more eco- and pedestrian-friendly for its dwellers and visitors. Applications such as Walk Score allow anyone interested in getting to know any American, Canadian and Australian city's walkability score and has expanded to record transit-friendly and bike-friendly cities in America.

In Asia, travel bloggers have voted their own fair share of their favourite walkable cities in Asia spanning from Malaysia to Mongolia, and Japan to Uzbekistan. Exploring a city by foot in Asia allows someone to bask in a city's local side amid thriving street culture, which can happen in some of the largest Asian metropolises. Planners are thinking hard on reshaping the urban scene to accommodate pedestrians' recreational needs. We should strive for greater efforts. We undoubtedly possess the potential to exceed our current achievements. Singapore's focus on walkability has helped to create a vibrant and dynamic city that is both sustainable and welcoming. We believe that Singapore offers a useful case study, a city that has made a deliberate effort over the years to transform its built environment to enable a greater degree of mobility through walking and cycling.

References

Asian Development Bank (ADB). (2011). *Walkability and pedestrian facilities in Asian cities state and issues*. Asian Development Bank (adb.org). Accessed on 22 August 2021.

Biewener, A. A., & Patek, S. (2003). *Animal locomotion*. Oxford University Press.

Caldwell, N. (2019, May 17). *Urban planners are saving the planet with redesigned, walkable cities*. https://www.greenmatters.com/travel/2018/05/23/22PR7j/urban-planners-cities. Accessed on 1 August 2021.

Cameron, R. W., Blanuša, T., Taylor, J. E., Salisbury, A., Halstead, A. J., Henricot, B., & Thompson, K. (2012). The domestic garden—Its contribution to urban green infrastructure. *Urban Forestry & Urban Greening, 11*(2), 129–137.

Cappello, N., & Buder, S. (2020, December 30). *No cars allowed: The 10 best pedestrian streets around the world*. https://www.afar.com/magazine/the-10-best-pedestrian-streets-around-the-world. Accessed on 1 August 2021.

Cavagna, G. A., Heglund, N. C., & Taylor, C. R. (1977). Mechanical work in terrestrial locomotion: Two basic mechanisms for minimizing energy expenditure. *American Journal of Physiology-Regulatory, Integrative and Comparative Physiology, 233*(5), 243–261.

Centre for Liveable Cities and Urban Land Institute. (2017). *Urban mobility: 10 cities leading the way in Asia-Pacific*. https://www.clc.gov.sg/research-publications/publications/books/view/urban-mobility-10-cities-leading-the-way-in-asia-pacific. Accessed on 7 April 2023.

Chang, C. Y., Woo, T. H., Wang, S. F., & Kuo, M.-J. (2010). *Survey and analysis of walking speeds for the elderly and children at crosswalks*. https://www.researchgate.net/publication/321913430_Survey_and_Analysis_of_Walking_Speeds_for_the Elderly_and_Children_at_Crosswalks. Accessed on 31 July 21.

Florida, R. (2002). *The rise of the creative class*. Basic Books.

Hanson, S., & Jones, A. (2015). Is there evidence that walking groups have health benefits? A systematic review and meta-analysis. *British Journal of Sports Medicine, 49*(11), 710–715.

Krambeck, H., & Shah, J. (2006). *The global walkability index: Talk the walk and walk the talk*. Paper presented at the Better Air Quality Conference (BAQ).

Levine, R. (1997). *A geography of time: The temporal misadventures of a social psychologist, or how every culture keeps time just a little bit differently*. Basic Books.

Lo, R. H. (2009). Walkability: What is it? *Journal of Urbanism, 2*(2), 145–166.

Mooney, C., & Freedman, A. (2021, January 25). Earth is now losing 1.2 trillion tons of ice each year. And it's going to get worse. *Washington Post*. https://www.washingtonpost.com/climate-environment/2021/01/25/ice-melt-quickens-greenland-glaciers/. Accessed on 14 January 2023.

Olafsdottir, G., Cloke, P., Schulz, A., Van Dyck, Z., Eysteinsson, T., Thorleifsdottir, B., & Vögele, C. (2020). Health benefits of walking in nature: A randomized controlled study under conditions of real-life stress. *Environment and Behavior, 52*(3), 248–274.

Roubini, N. (2022). *Megathreats: The ten threats that imperil our future, and how to survive them*. John Murray (Publishers).

Samra, H. A, & Specker, B. (2007). Walking age does not explain term versus preterm difference in bone geometry. *The Journal of pediatrics, 151*(1), 61–66. https://doi.org/10.1016%2Fj.jpeds.2007.02.033

CHAPTER 2

Singapore's Public Space Contest

Before the arrival of Stamford Raffles in Singapore in 1819, Singapore was already an important trading port. Archaeology excavations, uncovering ceramics, coins and beads, among other things, highlighted Singapore's prominence as a trading site and an entrepot, particularly in trading with China, India, Thailand, Vietnam and Indonesia. The Singapore port was governed by the Malay chieftains, then the Johor sultanate till it fell into the Portuguese hands in 1511, the Dutch in 1641 and the British in 1824 with the effect of the Anglo-Dutch Treaty. British arrivals in the nineteenth century strengthened Singapore as a central port of call in Southeast Asia. Its free port and central location attracted commercial activities between the European and Asian nations, contributing to the port's growth and Singapore's role in emerging regional trade. The opening of the Suez Canal in 1869 and new technology in communication and transportation brought rapid social and economic change, consolidating Singapore's position as the regional entrepot.

The expansion of Singapore's economy generated jobs stemming from the need for services to backbreaking work and the rest were serving the residents among their colonies. The able-bodied Chinese migrants pulled the rickshaws, an affordable form of transportation for the Europeans and Asians. The *chettiars* were moneylenders from Tamil Nadu and operated in warehouses or *kittangi* (in Tamil) along Market Street and Chulia

Melissa Liow Li Sa and Sam Choon-Yin, *Sustainable Urban Development in Singapore*, Sustainable Development Goals Series,
https://doi.org/10.1007/978-981-99-5451-3_2

Street which were close to the Singapore River, offering small loans to the Tamil and Chinese merchants and civil servants. Raffles was instrumental in homogenising the vision of Singapore, seeking to maintain dominance and colonial order over the servile local communities by classification, physical sanitisation and design according to social groups, races and the mercantile trade. Implicit rules of exclusion and monitoring boundary control through surveillance were formulated during the colonial era.

Philip Jackson's first Master Plan (also known as the Jackson Plan or Raffles Town Plan) under the watchful eyes of Raffles became the blueprint for Singapore's urban form. Published in 1822, the "Plan of the Town of Singapore" was drawn up to maintain some order in the urban development of the fledgling but thriving colony founded just three years earlier. The town plan was also designed to further the interest of British commerce in the region by allocating land and space according to the occupant's financial worth and function while seeking to maintain social order, boost business trade and safeguard the colony administration. The British cleared land to open up Singapore for cultivation, particularly for gambier and pepper plantations. Other crops cultivated include tapioca, indigo, fruits, vegetables and Liberian coffee.[1]

The central location was reserved for government offices, such as the chambers of government, the courts and police, and for the use of European and other merchants. Parallel roads of the Singapore coast heading north to the business trade entities and stretching alongside the river to an open field, known as "Padang" (field in Malay) became the primary developmental and business place. The town was partitioned into distinct areas based on ethics, provincial groups and nationalities such as European Town and China Town. The Europeans formed the upper social class and lived in the most urbanised part of Singapore. The Chinese quarter was stationed at what is today New Bridge Road and along the beach of Teluk Ayer on the assumption that the Chinese form an industrious and commercially active community in trading and other areas of commerce. The Malays were allowed to occupy areas around Kampung Glam whereas the Indians lived in the Serangoon Road vicinity where a whole lot of cattle farming and brick kiln trading activities were conducted there. The

[1] The concern over excessive logging and deforestation prompted the government to call on officials at the Singapore Botanic Garden (founded in 1859) to set aside land for a tree nursery to preserve exotic species such as ebony, teak and mahogany (see Barnard, 2016).

zoning process was not designed to promote equality, but rather to reinforce the fundamental divisions within colonial society, including those between the ruling and subject classes, as well as between Asians and Europeans and between the wealthy and the poor. "The imposition of the new paradigm of corridors, avenues straight lines and grids", Ooi Giok Ling argued, "was not only to create the image of imperial splendour or for pragmatics of municipal administration, but as a means of spatial sectorisation and insidious surveillance and thus control of the indigenous population".[2] There was inadequate housing built for the influx of immigrants who then stayed in overcrowded shop-houses in the expanding township or in fetid environments in native village huts along the waterways. The disease was widespread due to unhygienic practices such as using night-soil buckets to pick-up human waste.

Although the community was diverse, they were relatively acquiescent and accommodative with the space of the five-foot way expended to them. The five-foot way became the social inclusion space for the Asian community to run their daily affairs from trading, eking a living, hawking, sleeping and socialising round the clock. Inadvertently, invisible territories have begun to appear which were demarcated by different gangs of *Samseng* or the Chinese secret societies who sworn brotherhood to the Chinese underworld. It was mostly prevalent among the labouring classes who seek protection from their gangs in order to survive in the brutal environment. The *Samseng* gangs often fought hard for protection fees by terrorising street life. The local municipal authority seemed oblivious and did not pay much attention to the *Samsengs* since there was little municipal understanding of the *Samseng* activities. Another contributing factor was corruption. Corruption was prevalent during the British rule, especially among the European inspectors and the Malay and Indian rank and file.[3]

It was perhaps oversimplified to divide the private spaces for exclusive use by the owners in contrast to the poor governed public spaces which was open to all. The five-foot way was plagued with chaos, disease

[2] Ooi (2004, p. 58).

[3] In 1886, a Commission appointed to study public gambling in the Straits Settlements found police corruption to be very serious. For example, in the period between 1845 and 1921, a total of 172 cases of police corruption were reported in the local press with bribery being the most common form of police corruption (109 cases reported) followed by police involvements in thefts and robbery (42 cases) (Quah, 1979).

and sickness. A sanitisation programme to instil control was unequivocally implemented to address some of these issues. Harsh penalties, relocation exercises and court lawsuits were some of the measures undertaken by the municipal.[4] The local community resisted and did not cooperate as they perceived that the municipal wanted to tighten discipline, control and a demonstration of superintendence. The stronger imposition of the municipal on the classified space became a daily affair of contestation between the municipal and the Asian community, police and shopkeepers, as well as the *Samsengs* fighting with the coolies. This has resulted a sense of umbrage for the municipal and was persistent to clear the five-foot ways from debris and the festered activities accumulated by the heterogenous Asian community lifestyles.[5]

Independence in 1965

Singapore attained the rights to self-rule in 1959 from the British. Lee Kuan Yew became the first prime minister who led the People's Action Party (PAP) to win a majority of 43 seats in the first general election. Singapore joined the Federation of Malaysia in 1963, encompassing the North Borneo (presently Sabah) and Sarawak. The merger had to reshape in under two years, when Singapore left Malaysia and became a sovereign democratic country on 9 August 1965.

In 1965, Singapore was a typical third-world country. The city faced political uncertainty and urban decline. The roads were dirty. The city was overcrowded with thousands living in slums or squatter settlements without proper running water and sanitation. The government-led started a campaign to clean the city. People were educated not to litter. Laws were introduced to criminalise littering. Lee started the tree planting exercise to transform Singapore into a garden city. A large number of trees and shrubs have since been planted (roads leading to the airport were lined up with trees as Lee wanted foreign visitors to have a good first impression of Singapore as a clean and green city when they arrived). Efforts were put in to cleanup the Singapore River. Affected hawker stall owners were reallocated to hawker centres whereas residents living in nearby slums were

[4] For details, see Ho and Lim (1992).

[5] The contestation of space between the coloniser and colonised was a recurrent issue as demonstrated in Brenda Yeoh's (2003) *Contesting Space, Power Relations and the Urban Built Environment in Colonial Singapore.*

offered affordable public housing options. The city became cleaner over the years.

Developing Singapore into an economic hub to compete at the regional and international levels was a top priority. Without a sizeable domestic market, Singapore has no real economic base to operate from. Singapore is dependent on regional and global trade to ensure its economic viability. Singapore started with labour-intensive export-oriented processing industries because labour cost was low in the 1950s and 1960s. It moved up the technology ladder in the 1970s by attracting multinational corporations and shifting from labour intensive to knowledge-based economy.[6] By mid-1970s, Singapore had built good schools, affordable homes, factories and industry complex at Jurong to provide jobs for the people, parks and new townships, demonstrating her ability to plan to create better lives for the people. This was barely two decades after independence. This is all more important for the densest nation state in the world as supporting more people per square metre tends to be more complex and requires high standards in planning and urban design simple because towns are located very close to one another. Singaporeans have prospered. GDP per capita of Singapore has increased from around US$4000 in 1965 when Singapore became independent to more than US$55,000 today.[7]

New challenges surfaced in Singapore. As people and businesses prospered, private consumption increased. Singaporeans desired for larger houses and private vehicles to satisfy their aspirations, raising concerns about congestion and rising housing prices, especially in central Singapore. Urbanisation is inevitable. One of the strengths of Singapore is physical planning. The Concept Plan offers a broad, long-term use of land. The plan is reviewed once every 10 years, which is then translated

[6] This has enabled Singapore to enjoy robust growth in Gross Domestic Product (GDP) per capita. As Singapore moves closer to the technological frontier, the pace of technology diffusion and GDP growth rates slow down. See Hoon (2019).

[7] As reported in Fujii and Ray (2022).

to a more detailed plan and gazetted as the Master Plan.[8] The URA is a key institution to implement the plan.[9]

A significant milestone, after 25 years of independence, was the decision to establish new towns surrounding the central area to relocate the central area population. Singapore's commercial and housing needs are contained by ring-like self-contained settlements marked by nodal links via its train system and highways. The expansion of new towns, industrial estates and infrastructure, including transport services, led to a substantial increase in the built-up area. Specifically, the proportion of built-up land expanded from 26.5% in the late 1960s to 45.5% of the total land area by the early 1980s.[10] The measures helped to mitigate the overcrowding issues in Singapore's Central Business District (CBD). Never complacent with past achievements, the design of the towns in the suburban areas is constantly upgraded to offer comfort and convenience. The focus was to ensure that the facilities and places were convenient (consisting of office blocks, shops, cinemas, malls, schools and green spaces) and accessible (integrated with the public transportation system and well connected by a system of expressway and roads and MRT). The area allocated to parks and gardens also increased significantly to reflect "the government's political astute realisation that, as Singaporeans move up the economic ladder and demand a better quality of life, they would want to enjoy good recreational grounds and live in pleasant and aesthetically pleasing environments".[11] Part of the design is also to make Singapore more pedestrian friendly and integrated with the public transportation system.

[8] The Concept Plan takes on a more dynamic view in planning for the physical development of Singapore. The Concept Plan is an official statement of a geographic unit's policies and intentions pertaining to physical development and establishes the framework to guide government agencies in their effort to assemble and develop land for public purposes. It provides general directions as to where, what, when and why with regard to all recommendations set forth within the plan. The Master Plan, on the other hand, is a statutory instrument meant to control the use and development of land use in Singapore through zoning, density and plot ratio ceilings. The Plan specifies the allocation of land for uses such as schools, open spaces, infrastructure facilities and other essential needs of the community.

[9] Established in 1974, the URA coordinates the redevelopment plans, including conservation of heritage buildings, shophouses and neighbourhoods.

[10] Ooi and Kog (1999, p. 178).

[11] Savage (1997, p. 189).

The MRT transport system was launched in 1987 with five stations from Toa Payoh to Yio Chu Kang. Today, more than 140 stations across six MRT lines span the island, offering accessibility for HDB residents to travel to the CBD area. Coupled with the subsumption of the retail, the development enables continuum of public and private space which allows for some form of reclamation of the innercity by the previously evicted locals, while encouraging movement, co-presence and recreation that connects the spatial and transpatial groupings. During the weekdays, the MRT services enhance the commuters' accessibility to the central area. In the weekends, they offer an affordable form of transportation to recreational destination to transient, spatial and transpatial groupings. Dhoby Ghaut MRT Station is one of the busiest underground interchange stations located in the eastern end of Orchard Road shopping area, linking the Circle Line, Northeast Line and North South Line. It has a street-level connection to The Atrium @ Orchard shopping mall, commercial areas, education institutions, residential apartments, museums, places of worship and the Istana nearby. A green space can be found at its ground level, adding greenery to the compact-built Orchard Road belt.

All these would not have been feasible if the environmental quality is poor. Singapore understood that it had to take into consideration the adverse effect of industrialisation on the quality of environment. In April 1970, the Anti-Pollution Unit (APU) was formed. Established under the Prime Minister office, the APU has regulatory power to reject industries deemed to be too polluted from operating in Singapore. The Clean Air Act was passed in 1971 to prevent and control air pollution. The stringency in effluent and emission regulations was deemed necessary as local companies tend to view environmental protection as a deterrent to profit generation. The Ministry of Environment was later instituted in 1972 and took over the task of cleaning up the Singapore River and Kallang Basin Catchment. Today, a variety of restaurants and retailers operate around the river, adding jobs and tourists. Streets and lanes spruced up, many of them repurposed as space for walking for the locals and tourists. Cleaning of the Singapore River depicts a case of Singapore's effort to remake its city centre.

Despite the presence of diverse mixed-use environments, it is important to note that transport and retail infrastructure schemes create classified environments with implicit and explicit rules of exclusion or inclusion. The rules are mostly implicit with the retail outlets located within the concourses that appeal or are less appealing for others during different times of the day. The explicit rule, on the other hand, demarcates physical boundaries which invisibly prevent those who are not allowed to travel to other platforms. The ability of the concourse to offer an interface and connection with the sophisticated retail environments does bear similarity to the colonial five-foot way. The difference in the post-colonial era is that the classification has become more robust as compared to its predecessor due mainly to the clear maintenance of boundaries and rules of governance. With the connecting space which authorises movement as well as the explicit rules on what is permissible or not when using the spaces, any obscurity would be eliminated with the installation and capture of closed-circuit television (CCTV) and security guard presence. The explicit rules secure the boundaries of the dominant power or corporations and limit the space contest between the corporation and store or office tenants, often nullifying any emerging contests. Its weaker classification of the transient, spatial and transpatial groupings offers movement, co-presence and recreation opportunities and the rising corporate dominance has socially sanitised the privatised public space. With closer scrutiny, the contest for space is often rescinded, allowing a prevailing harmonious society for Singapore.

Not all had turned out the way that the state wanted. Amid the homogenous and corresponding urban design that carried on from the colonial and post-colonial times, the five-foot way shop-houses and the void decks had defined weaker classifications, causing contestation of space among the residents, as well as the municipal and the residents. Greater social rules were thrusted with the strengthening of the spatial framing and classification in the interests of efficiency and modernity. On the flipside, non-corresponding designs of inclusion of transpatial and spatial groupings were formed due to the increased surveillance and intensified rules of exclusion might lead to under-utilisation of space of the five-foot way shop-houses and HDB void decks.

Urban development effort was transpired to make way to appeal the international and transpatial social groupings in the CBD area, and a huge relocation exercise of local spatial groupings to the HDB blocks located in the CBD periphery became the state's top priority during the post-colonial era. The municipal realised that in their pursuit of urban redevelopment, a large number of heritage shop-houses along the quayside were destroyed completely.

A non-correspondent planning method appeared in tandem with the MRT system to ease movement for locals from the HDB blocks to the

CBD, enabling locals to reclaim some space and integrate with transpatial groupings and the transient tourists visiting Singapore. However, it remains that stringent considerations and difficult trade-offs among multiple objectives must be made. Allocating space for pedestrians could lead to less space for commercial development. In regard to the location of the bus interchange, Lim Leong Geok, former planner and rail expert with Singapore Mass Rapid Transit, said in an oral interview that the HDB was asked to build a bus interchange in the town centre especially if there is an MRT station. URA had wanted the bus interchange located next to the MRT station, the Housing and Development Board was "not very supportive". But HDB said no as it wanted the space next to the station for commercial development. In Junction 8, Lim said, "the MRT is there and in between, you've got this commercial development. The bus interchange is somewhere else. So, there's a long walk for people". It took former Chief Planner and CEO of the URA Liew Thai Ker to change the direction who went further to suggest building a bus interchange underground to be of proximity to an underground MRT.[12]

It is worth noting that while land area in Singapore has increased over the years through land reclamation, Singapore remains a physically small country.[13] The implication is straightforward. Any developmental plan will have to balance economic, social and environmental interests. Delicate care was undertaken with the help of the United Nations experts in the early years of independence to preserve and rehabilitate the colonial two to three-storey shop-houses to maintain the privatisation of public space and encourage social integration, with the hope to enhance the connection between organisations and the civil people. But some of the shop-houses had to be destroyed to make way for a new podium and tower form during the intense urbanisation era. Indeed, tension arose when it comes to land allocation. Conservationists have raised concerns about the loss of the country's biodiversity when land is cleared for urban

[12] Lim Leong Geok (accession No. E000015), Oran Interview on March 16, 2004, Oral History Centre, National Archives of Singapore https://www.nas.gov.sg/archivesonline/oral_history_interviews/record-details/d1509ee5-1161-11e3-83d5-0050568939ad (Accessed on 16 December 2022).

[13] Examples of land reclamation projects include the Kallang River in the 1960s for industrial and housing purpose, the Jurong Island in the 1990s which houses the petrochemical complex, and the 38 hectares *of land reclaimed* at Bayfront in 1994 to create the shore profile *of Marina* Bay.

development purposes. The debate between nature preservation requirements and the state's argument to utilise the land for purposes other than nature preservation will continue for years to come.

Similarly, it can be argued that Singapore's industrial policy has been primarily oriented towards emphasising the economic aspect. The case of Jurong Island is worth noting. Opened in October 2000, Jurong Island houses the petrochemical complex and is home to many companies, including Shell, ExxonMobile, DuPont and Sumitomo Chemicals.[14] The EDB felt that the Jurong Island project was worth pursuing as it was expected to generate high-paying jobs and attract massive foreign direct investments into Singapore. However, the industry has an adverse effect on the environment because it makes use of significant levels of non-renewable fossil resources (oil and natural gas) in addition to the cutting down of trees to make way for the complex and its infrastructure. This does not mean that there was a lack of deterrence measures against polluters.[15] But the introduction and expansion of the petrochemical industry automatically raise the high emission levels of greenhouse gasses in the city-state because of the significant use of non-renewable fossil resources (oil and natural gas).

While cracks appear from time to time as Singapore strives to balance social, economic and environmental interests in its urban design, Singapore's experience has won praise from some observers after making comparisons with other countries. Former Foreign Minister George Yeo illustrates the case with reference to Hong Kong. Hong Kong has a total land area of 1100 square kilometres of which as Yeo noted 800 square kilometres, which is larger than the whole of Singapore, are green zone and reserved for agriculture activities and parks. "Unlike Singapore, Hong Kong has no need for air and naval bases, power stations or large water reservoirs. The land shortage in Hong Kong is thus completely artificial, partly created to keep the value of land high. The government itself depends heavily on land revenue. The unfortunate result is that many

[14] Seven islands were merged to form Jurong Island in the western part of Singapore to provide a location for the petrochemical complex.

[15] The Environmental Protection and Management Act stipulates that any premises that cause pollution of the environment or are injurious to public health or safety are "guilty of an offence and shall be liable on conviction to a fine not exceeding S$100,000 or to imprisonment for a term not exceeding 3 months or to both and, in the case of a continuing offence, to a further fine not exceeding S$2000 for every day or part thereof during which the offence continues after conviction" [Chapter 94A, Act 39(2)].

Hong Kongers have no hope of being able to own a decent roof over this head after working their entire lives".[16]

Inevitably, the government has a leading role in urban planning and development. Streets, other public spaces and land use are involved. The transformation of the city to encourage walkability requires a strong role of the government to allocate land for competing use such as housing, commerce, transport, parks and recreation and a profound sense of duty and positive behaviour across the public sector. To write about any paradigm to make cities better in this regard, we cannot ignore the role of government nor the legitimacy of the government's actions and plans for the future of cities. In the case of Singapore, some scholars have described Singapore's political system as illiberal democratic,[17] authoritarian capitalism,[18] non-competitive, semi-democratic[19] and soft authoritarian.[20] Cherian George coins the term "calibrated coercion" to describe Singapore's approach to managing the media. The approach entails behind the scene controls that create the conditions for self-censorship, for example, through the Newspaper and Printing Presses Act 1974, which gives the government the controlling power to issue licences to companies to operate in the city-state, ensuring that "newspaper companies would focus on commercial success" and align "their interest with a pro-growth government".[21]

Social progress and economic success in Singapore, writes N. C. Saxena, has been "strongly driven by a government which was heavily involved in every area of national development, operating through effective and highly competent public institutions which are deemed to be among the least corrupt in the world".[22] Edgar Schein has labelled Singapore's development strategy as "strategic pragmatism" in the sense that the government "was very clear about its long-range goals but, at the same time, remained tactically very flexible in working towards these goals

[16] Yeo (2022, p. 424).

[17] Mutalib (2000).

[18] Lingle (1996).

[19] Case (1996).

[20] Means (1996).

[21] George (2012, p. 99).

[22] Saxena (2011, p. 2).

and nimble in solving the day-to-day problems".[23] Pragmatism can be simply translated as "being practical", and practice policies of the PAP government have been able to deliver the goods.

While the predominance of a single political party has its shortcomings, it allows for a competent government to focus its energy on development goals that matter to the people and be far more involved in administration than political fights, which may have contributed to the PAP's legitimacy to govern. "The massive improvement in the material life of the entire nation", argued Chua Beng Huat "makes it very difficult to argue against success. It gives the (PAP) government a very high degree of political legitimacy, and the government is unlikely not to remind the electorate of its success in providing collective consumption goods and services".[24] Essentially, the voters were able to experience a system that had benefited them and were ready to keep the PAP in power instead of replacing it with a party that might not be able to produce similar or better results than the PAP.

Based on the election results for the past fifty years there is unquestionable strong support for the PAP. Singaporeans are attracted to the party's ability to bring security and wealth to them. Singapore has achieved top ranks in the efficiency of the labour and financial markers, level of competitiveness and infrastructure. Singapore residents enjoy first-class education and healthcare standards, and more than 90% of them own high-rise public housing. It ranks among the top in development indicators such as life expectancy and infant mortality rate. The water and environment are clean. In Singapore, writes Kishore Mahbubani "there are no homeless, destitute, or starving people... Poverty has been eradicated... The state acts as the catalyst—matching financial support, sponsoring preventive and social care, and ensuring that basic needs are provided for. Remarkably, the poorest 5% of households have about the same levels of ownership of homes, television sets, refrigerators, telephones, washing machines and video recorders as the national average".[25]

A necessary condition to gain the people's support for state-led model of economic development is good governance—to provide the assurance to the public that the economic gains are effectively utilised to generate

[23] Schein (1996, p. 23).

[24] Chua (1997, p. 313).

[25] Mahbubani (2015, p. 58).

greater social good, and not expropriated by the politicians. In the case of Singapore, the World Bank's "Governance Matters" report, which assesses the standard of governance of more than 200 countries, has rated the city-state highly. Singapore obtained a score of close to 100 for five of the six categories of the quality of governance. In terms of corruption in the public sector, international agencies have consistently ranked Singapore as one of the least corrupt countries in the world. Transparency International and Hong Kong-based Political & Economic Risk Consultancy (PERC) Limited have consistently ranked Singapore as one of the last corrupt countries in the world. When Singapore commenced self-control in 1959, the PAP made it a national priority to combat corruption. The ruling party recognised the importance of making Singapore corruption-free both in the government and business environments in its attempt to develop the country's economy. As the then Prime Minister Lee Kuan Yew said:

> All new governments want to prove themselves by passing many new laws and launching many new projects. We hit the ground running, before the phrase was coined. ... Most important was a bill to give ourselves (the PAP) wider powers to combat corruption. It was the first of several that strengthened the law so that offenders could be charged and convicted in court. It led to the creation of a new agency, the Corrupt Practices Investigation Bureau (CPIB), which has helped to keep Singapore clean.[26]

Political legitimacy is about the government's ability to appeal to the masses by doing what is right for the nation. Singapore's ability to transform the city—despite the challenges—suggests a high level of trust and support the Singapore government has received from Singaporeans, the belief that the government is clean and effective, and has the capacity to protect the interest of the people.

It is evident that developments over the last three decades have seen more mixed-use projects that attracted a broader spectrum of society; the purposely built-in heterogeneity of public space became a less contested domain. By establishing connections and fostering collaborations with various stakeholders, including civic society and think tanks, the strategic perspective is that each stakeholder has unique and essential roles in collaboratively shaping and innovating the city-state.

[26] Lee (1998, p. 346).

Singapore has been proactive in its efforts to prepare for the impact of the Fourth Industrial Revolution, recognising the need to adapt and innovate in order to thrive in this rapidly changing landscape. Recognised as a Smart Nation, Singapore city is known for being an innovation-driven country. Government agencies spearheading the nation's intelligent transport system (ITS) programmes, for example, gather end-user needs and ground reactions in their bid to fulfil the ITS Smart Mobility 2030 goals. An integral role in policing public transport is to integrate and use ITS technology to formulate a sustainable transportation system for the business community, residents and inbound visitors to Singapore. While research institutions help to deliver technological innovations in transportation, the government and other key industry players play an instrumental role as they possess the financial strength to invest in location-based technologies, sensors and in-vehicle telematics, which serve as the primary drivers in Singapore's pursuit of the Smart Mobility 2030 programme. The strengthening of the local ITS capabilities enables the Singapore city in becoming a prime test-hub and transportation research area while gearing towards addressing the public aspiration for more open spaces to choose active mobility as a form of transport. This topic will be discussed in Chapter 9. Let's first explore how Singapore actively promotes walkability.

References

Barnard, T. P. (2016). *Nature's colony: Empire, nation and environment in the Singapore Botanic Gardens*. NUS Press.

Case, W. (1996). Can the halfway house stand? *Comparative Politics, 28*(4), 437–464.

Chua, B. H., (1997). Not depoliticized but ideologically successful: The public housing programme in Singapore. In J. H. Ong, C. K. Tong, & E. S. Tan (Eds.), *Understanding Singapore society*. Times Academic Press.

Fujii, T., & Ray, R. (2022) Singapore as a sustainability city: Past, present, and the future. In H. T. Hoon (Eds.), *The Singapore economy: Dynamism and inclusion*. Routledge.

George, C. (2012). *Freedom from the press: Journalism and state power in Singapore*. NUS Press.

Ho, K. C., & Lim, N. E. V. (1992). *Backlanes as contested regions: Construction and control of physical space*. Singapore University Press.

Hoon, H. T. (2019). *Economic dynamism, openness, and inclusion: How Singapore can make the transition from an era of catch-up growth to life in a mature economy*. World Scientific Publishing Corporation. https://doi.org/10.1142/10883

Lee, K. Y. (1998). *The Singapore Story*. Singapore Press Holdings and Times Edition.

Lingle, C. (1996). *Singapore's authoritarian capitalism: Asian values, free market illusions, and political dependency*. Edicions Sirocco, S.L.

Mahbubani, K. (2015). *Can Singapore survive?* Straits Times Press.

Means, G. (1996). Soft authoritarianism in Malaysia and Singapore. *Journal of Democracy, 7*(4), 103–117.

Mutalib, H. (2000). Illiberal democracy and the future of opposition in Singapore. *Third World Quarterly, 21*(2), 313–342.

Ooi, G. L. (2004). *Future of Space*. Eastern University Press.

Ooi, G. L., & Kog, Y. C. (1999). Future Urbanisation. In Low, L (Eds.), *Singapore: Towards a developed status*. Centre for Advanced Studies and Oxford University Press.

Quah, J. (1979). Police corruption in Singapore: An analysis of its forms, extent and causes. *Singapore Police Journal, 10*, 7–43.

Savage, V. (1997). Singapore's garden city: Translating Environmental possibilism. In G. L. Ooi, G. L., & K. Kowk (Eds.), *City and the state: Singapore's Built environment revisited*. The Institute of Policy Studies and Oxford University Press.

Saxena, N. C. (2011). *Virtuous cycles: The Singapore public service and national development*. United Nations Development Programme.

Schein, E. H. (1996). *Strategic pragmatism*. The MIT Press.

Yeo, G. (2022). *Musings (Series One)*. World Scientific.

Yeoh, B. S. A. (2003). *Contesting space in Colonial Singapore: Power relations and the urban built environment*. Singapore University Press.

CHAPTER 3

Walkability in Singapore

In Chapter 2, we pointed out that Singapore was a typical third-world country when independence was thrusted upon her. The roads were dirty. The city was overcrowded with thousands living in slums or squatter settlements without proper running water or sanitation. Economic development was the main driver of government policies. Sustainable development took a backseat.

In more recent years, Singapore has significantly prioritised sustainable development across all policy domains, recognising the importance of fostering a holistic and balanced approach to enhance the overall human experience. As a low island in the tropics, Singapore understands that without due care for the environment and consideration for the effect on food and water supply resulting from climate change, the worsening of the environment can impose a severe impact on the city-state. Among other things, Singapore seeks to reduce traffic congestion and induce people to shop and work by taking the bus or a short walk or cycling from where they live. Construction of cycling paths and park connectors forms a core strategy in making Singapore greener and more sustainable to contribute to Singapore's overall goal of reaching net-zero carbon emissions by 2050. It is appropriate that we emphasise Singapore's initiatives in advancing sustainability.

Melissa Liow Li Sa and Sam Choon-Yin, *Sustainable Urban Development in Singapore*, Sustainable Development Goals Series, https://doi.org/10.1007/978-981-99-5451-3_3

In this chapter, we will examine Singapore's sustainability efforts, and explore Singapore's efforts to promote walking and cycling and how they are shaping the mindset of pedestrians and cyclists.

Singapore's Sustainability Efforts: An Overview

Singapore is a small island developing state, sharing many of the characteristics of the Pacific Islands including dense population and small physical space. Although the city-state is largely protected from the hurricane and other natural disasters, rising sea level is a matter of concern. As Ng Wei-Shiuen and Robert Mendelsohn have shown, sea level inundation could lead to the loss of dry land of up to 17 square kilometres (2.7% of the Singapore's total land area).[1]

Without any significant agricultural land or landfill sites, Singapore does not emit significant amounts of methane gas. Carbon dioxide (CO_2) is the primary greenhouse gas emitted in the city-state, resulting from the use of fossil fuels in power generation and the industrial, transport and commercial and residential sectors.[2] In 2020, Singapore ranks 126th of 142 countries in terms of CO_2 emissions per dollar GDP. Singapore ranks 27th out of 142 countries in terms of emissions per capita.[3] Globally, Singapore accounts for only 0.2% of global CO_2 emission. Hence, Singapore's decision whether to support the international call for the reduction of greenhouse gasses would have no significant consequences

[1] Ng and Mendelsohn (2005).

[2] CO_2 receives greater attention because of its more globalised effects. The Kyoto Protocol has set up a global CO_2 market to reduce greenhouse gas emissions. Singapore is a signatory to the United Nations Framework Convention on Climate Change (UNFCCC). Singapore ratified the UNFCCC in 1997 and acceded to the Kyoto Protocol to the UNFCCC in April 2006. In 2009, Singapore pledged to reduce emissions by 16% from business-as-usual (BAU) levels by 2020. Singapore's decision to accede to the Kyoto Protocol in April 2006 has been described as nationalistically orientated "to protect the country's international image and to benefit from the economic opportunities that the protocol opens up". It represents a significant shift in Singapore's policy towards CO_2 emission, from denying its obligations to set emission targets as it clings on to its developing country status to the announcement of putting in place a plan to reduce carbon intensity.

[3] As reported by Singapore's National Climate Change Secretariat. https://www.nccs.gov.sg/singapores-climate-action/singapore-emissions-profile/. Accessed on 25 January 2023.

on greenhouse gas emissions in *global* terms. In fact, Singapore's position is that it could not volunteer to reduce its emissions at the cost of economic growth if other countries fail to do their part. At a student forum in October 2009, Lee Kuan Yew resisted attempts to list Singapore as an Annex 1 country at the international summit on climate change in Copenhagen, which would require the city-state to reduce greenhouse gas emissions by 50 to 85% by 2050. Lee pointed out that Singapore relied on the manufacturing sector to boost its economy, and hence complying with the condition would create adverse effects on its economic growth.[4] While Singapore committed to cut greenhouse gas emissions by 16% by 2020 in the run-up to the 2009 Copenhagen Summit, Singapore was clear that this would take effect only if all countries cut emissions and if other countries offer significant pledges.[5] A month before the summit in Copenhagen, the government of Singapore reaffirmed its position to reject emission cuts on the basis that the city-state's carbon dioxide emission is "negligible" as a percentage of global output and that it is "not obligated" to set targets to reduce emissions'.[6] The meeting at Copenhagen ended with an accord that contained no specific targets for emission cuts and non-legally binding commitments to limit global warming to 2C or less, as well as a promise of US$ 30 billion over three years to fund mitigation and adaptation efforts.

The "Copenhagen Accord" was brokered by five nations—United States, China, India, Brazil and South Africa—effectively snubbing the remaining 187 nations in a last-minute attempt to avoid a total collapse. It is worth noting that despite knowing the challenges that make negotiation difficult, Singapore was keen to work with others at Copenhagen to bring about an agreement as indicated by its commitment to cut 16% in greenhouse gas emission. Singapore was one of only 19 major economies that eventually committed to the Copenhagen Accord, which, if achieved, would result in global emission reductions of 9.1% from the business-as-usual level but still an increase of 11% above 2007 levels. After the accord was presented, Singapore was quick to issue a statement to show

[4] Ho (2009, October 30).

[5] Singapore's commitment pales in comparison with the 40–50% cut committed by China but aligns closely with the United States declaration to reduce US greenhouse gas emissions by 17% of 2005 levels by 2050 under the Waxman-Markey American Clean Energy and Security Act of 2009. Cheam (2009a, December 3).

[6] Gunasingham (2009, October 30) and Nur Dianah Suhaimi (2009, October 31).

its support for the draft. The statement read that Singapore "supports the sentiment, commitment and above all good faith in which it has been worked on and presented now to all parties, and we lend our voice to that of our colleagues in the Association of Small Island States and others who have called for it to be the basis to move us forward in Copenhagen and beyond".[7]

Singapore is also a signatory of the Paris Agreement. The Paris Agreement of 2015 was committed to limit warming to "well below 2 degrees Celsius above pre-industrial levels and pursuing efforts to limit the temperature increase to 1.5 degrees Celsius ...". Unlike the Kyoto Protocol which required only developed countries to take on emissions targets, the Paris Agreement was adopted by 196 parties. The Paris Agreement entered into force on 4 November 2016. Singapore signed the Paris Agreement on 22 April 2016 and ratified it on 21 September 2016. Singapore pledged to reduce its Emissions Intensity (EI)—the amount of GHGs emitted per dollar GDP—by 36% from 2005 levels by 2030 and to stabilise emissions with the aim of peaking around 2030.[8] On 31 March 2020, Singapore submitted its enhanced Nationally Determined Contribution (NDC) and Long-Term Low-Emissions Development Strategy (LEDS) document to the United Nations Framework Convention on Climate Change (UNFCCC). A key thrust in enabling Singapore's transition to low-carbon economy is the transformations in industry, economy and society to allow for more renewable energy usage, greater energy efficiency and reduction in energy consumption.[9]

[7] See "We are disappointed so far: PM" *The Straits Times*, 22 December 2009.

[8] The Climate Action Plan in 2016, which outlines Singapore's strategy for reducing greenhouse gas emissions and adapting to the impacts of climate change. The plan set a target to reduce the country's emissions intensity by 36% from 2005 levels by 2030, and to stabilise emissions with the aim of peaking around 2030. The plan also includes measures to increase energy efficiency, promote the use of renewable energy, and encourage sustainable transportation. In 2019, The Climate Action Plan was updated to Climate Action Plan 2030, which includes the nation's plan to become carbon neutral by 2050. The plan set a target to reduce the country's emissions intensity by 36% from 2005 levels by 2030, and to stabilise emissions with the aim of peaking around 2030. The plan also includes measures to increase energy efficiency, promote the use of renewable energy, and encourage sustainable transportation.

[9] While Singapore wants to contribute to the success of international meetings rather than be a spoiler and block the talks by attaching conditions, Singapore's strategy also implicitly claims the right to development while rejecting emission reduction if this might adversely affect its rate of economic growth. Any carbon emissions reduction, as the

Singapore has made several contributions to the United Nations' efforts to ensure environmental sustainability, as part of the United Nations Millennium Development Goals, particularly in the areas of waste management, energy efficiency and sustainable transportation. Singapore has implemented a comprehensive waste management system, including recycling and waste-to-energy initiatives, which has helped to reduce the amount of waste going to landfills. Policies and programmes to promote energy efficiency, such as the Energy Conservation Act and the Green Building Masterplan have helped to reduce energy consumption and emissions. On sustainable transportation, Singapore develops an extensive public transportation system, including buses, trains and bike-sharing programmes to reduce dependence on cars and decrease traffic congestion and air pollution.

At home, the Singapore Green Plan (first launched in 1992 with the latest version being Singapore Green Plan 2022) outlines Singapore's long-term plan for enhancing the resilience and sustainability of Singapore's environment and infrastructure, focusing mainly on three main areas: greening the urban environment, improving air and water quality, and reducing the impact of climate change. The plan includes targets to increase the proportion of green cover in Singapore, and to improve

Minister for the Environment and Water Resources noted, cannot "compromise economic growth" as doing so would be akin to "shooting ourselves in the foot". Cheam (2009b, December 3).

air quality through the reduction of emissions from transportation and industry.[10]

In 2014, to take another example, the Sustainable Singapore Blueprint was unveiled. The blueprint set targets for sustainable development in areas such as water and energy efficiency, waste management and biodiversity conservation. It aims to create a more liveable and sustainable Singapore by 2030, through the implementation of sustainable urban solutions in areas such as transportation, housing and water management. The blueprint also included targets to reduce carbon emissions and improve air quality, as well as increase the proportion of green cover in Singapore.

An essential element of sustainability which deserves greater attention is walkability. Walkability forms a crucial part of Singapore's sustainability efforts as it reduces the reliance on personal vehicles, which in turn decreases the urban heat island effect and the amount of energy needed to power transportation. By prioritising walkability, Singapore is making progress towards achieving these goals and creating a more sustainable and liveable city. How does Singapore encourage people to walk?

[10] The plan includes specific initiatives and targets such as:

- Increasing the proportion of green cover in Singapore to 35% by 2030.
- Improving water quality in Singapore's reservoirs and waterways.
- Reducing the carbon footprint of buildings.
- Improving the energy efficiency of buildings and vehicles.
- Promoting sustainable transportation, such as cycling and electric vehicles.
- Enhancing biodiversity and protecting endangered species in Singapore.
- Encouraging sustainable consumption and production.

Walking in Singapore

Walking as well as cycling offer Singapore a way to deal with the urban transportation challenges and simultaneously promoting sustainability, health and quality of life in the city-state. The "Walk Cycle Ride" programme is the key campaign to promote walking and cycling in Singapore. Pathways are constructed along major roadways, many of which are sheltered to provide greater comfort to the users. Pedestrian crossings and overhead bridges are common in Singapore. Signages to enhance the safety of pedestrians and cyclists are visible.

Much effort has been put into expanding the public transport system such as increasing the number of trains, upgrading the signalling system of the rail lines to allow for more trains to run thereby reducing overcrowding and shortening of waiting time, and extending the rail network to include more train stations and rail lines. The use of public transport to meet the first/last-mile stages of trips from stations to home and vice versa provides the benefits for users to walk and cycle. One can walk or cycle from home to the train station and proceed to the destination with the train services. This way, travelling time is saved.

Overall, residents in Singapore demonstrate adaptability and embrace walking as a convenient means of transportation for accessing various destinations within their neighbourhood on foot. Walking is perceived as not needing too much help and effort from others, and that is why the independence of this mode of transport has been popular even among the elderly population in Singapore. Rojas Lopez and Wong Yiik-Diew conducted a study that highlights the widespread use of walking as a mode of transportation among adults. Walking serves as a common choice for first/last-mile travel, as well as for commuting to work or school. The research also shows that among the elderly population, walking is primarily utilised for first/last-mile travel, with shopping or running errands being the frequent reason for walking.[11]

Ensuring mobility and accessibility is not a given. Key considerations for walkability include prioritising safety, comfort and convenience. Singapore's high temperatures can serve as a deterrent for individuals who are considering walking as a mode of transportation. The scorching heat can

[11] Lopez and Wong (2017).

make walking uncomfortable and even unpleasant.[12] Young schooling children may be less willing to walk far under scorching sun and heavy rain as they must carry their heavy school bags to schools and tuition centres. Tourists with heavy backpacks and luggage would face similar circumstances.

Koh Puay Ping and Wong Yiik Dew surveyed people at nine MRT station exits on 12 factors potentially influencing their decision to walk.[13] The presence of rain shelters, the number of road crossings/delay, comfort, traffic risk and directional signs were ranked more highly while scenery and availability of shops along the route were less important. Piotr Olszewski and Sony Sulaksono Wibowo investigated walk access to MRT by interviewing people at 11 MRT stations.[14] They found rain shelters, comfort (number of steps to climb), security, traffic accident risk and number of road crossings as important factors.

To improve road safety for pedestrians, some 200 kilometres of sheltered walkways have been added under the "Walk Cycle Ride" programme nationwide.[15] A total of 50 Silver Zones in selected residential spaces has been planned up till 2023 to improve road safety for pedestrians, especially the elderly. Some of the road safety innovations include (1) yellow rumble signs and bright fluorescent yellow-green signs to signal motorists that they are close to a Silver Zone, (2) rest points along the road dividers, (3) chicanes and rumble stripes with reduced width lanes to keep a low vehicle speed of 40 kilometres per hour, (4) low height centre dividers to motivate motorists to drive slow and (5) roundabouts to smoothen traffic conflict points and traffic flow.

The emphasis on walkability is a shift away from the focus on making roads more efficient for vehicular movement as this translate to repurposing some parts of the road, including cutting lanes to make way for footpaths. The transformation of Singapore streets enables pedestrians and cyclists to walk and cycle for work in addition to those who participate in these alternative modes of transport for leisure and recreational purposes, which are already served by the park connectors. A noteworthy

[12] Temperature in Singapore may reach 33 degree Celsius in the day and humidity level of 100% during prolonged raining seasons with mean annual relative humidity of about 84%. Direct irradiance from sunshine is about 120 Watts/m^2.

[13] Koh and Wong (2013).

[14] Olszewski and Wibowo (2005).

[15] Government of Singapore (2022).

project by the URA is the transformation of Singapore's *Padang* into a civic open space in the heart of the city district. URA works with the LTA to reclaim roads such as parts of Fullerton Road. Empress Lawn is carved from realigning' between the Anderson Bridge and the Singapore Cricket Club, which now locates in front of the Concert Hall and Victoria Theatre. Improving the landscape along the reclaimed road spaces curates a safe and appealing green oasis for visitors on foot. In November 2021, the LTA announced the closure of Connaught Drive and Anderson Bridge in the Civic District area, the birthplace of modern Singapore, to allow pedestrians to walk more easily between the Asian Civilisation Museum, Victoria Theatre and Concert Hall, National Art Gallery and the Arts House, offering pedestrians "a green, idyllic and elegant cultural precinct where they can walk in leisure to enjoy a glimpse of the past without having to compete with vehicular traffic", said LTA's Chief Executive Ng Lang. Mature trees are planted to provide shade. The air is cooled by the breeze from Marina Bay. This is part of Singapore's plan to turn the area around the *Padang* into a walkable park announced by then National Development Minister Khaw Boon Wan in 2014.

There has been a concerted effort to bring greenery and nature into public spaces to create a more liveable environment to realise the "garden city" vision first put forward by then Prime Minister Lee Kuan Yew in 1967. One notable example is the Tengah, or the "Forest Town", which upon completion, will be the first smart and sustainable town, housing more than 40,000 new homes and featuring a car-lite city centre with walking and cycling paths along the roads, greenery woven around the town and use of smart technologies to, among other things, optimise energy consumption and automate waste collection. The forest town preserves a 328-foot-wide ecological corridor where its centre links the water catchment area to the nature reserve. Its smart energy management system and Pneumatic Waste Conveyance System (PWCS) (a high-speed air to transport waste to a centralised bin centre) are sustainable features of the forest town that further cement Singapore's pledge to become a smart city. Residents can visit other towns or the city through buses, and with roads running beneath the town centre, it frees up spaces on the surface, making it more friendly for cyclists and pedestrians. The rationale for integrating the natural systems is to improve biodiversity, reduce the urban heat inland effect thereby improving the level of comfort of outdoor activities. Speaking at the 2012 World Cities Summit, Prime Minister Lee Hsien Looing told his audience: "In the next phase, our aim is to build a 'City in a Garden', to bring green spaces and bio-diversity to our doorsteps."[16]

Bukit Panjang, to take another example, is a town in the western part of Singapore that has been designed to be pedestrian and cycling friendly. The town boasts a comprehensive network of cycling paths and park connectors, spanning a total distance of 16 kilometres, with almost every street having a designated cycling path. This allows residents to easily walk or cycle to community centres, hawker centres and shopping malls. Additionally, the cycling paths are connected to train stations, which enhances fast- and last-mile connectivity as well we to neighbouring towns such as Choa Chu Kang and the Rail Corridor. The train stations are equipped with bicycle parking facilities for added convenience. The Rail Corridor, which spans 21 kilometres and is one of the five corridors, is a walking trail that stretches from Kranji MRT station in the north to Tanjong Pagar in the south.

The Singapore government recognises the importance of educating and creating awareness among the public about the benefits of walking. As part of its efforts, it runs various public awareness campaigns through different media channels, such as TV commercials, billboards and social media platforms. The campaigns aim to promote the benefits of walking and encourage people to adopt walking as a viable mode of transportation. For instance, the "Car-lite Singapore" campaign was launched in 2016 to encourage the public to use alternative modes of transport, including walking, and reduce their dependency on cars. The campaign included a series of roadshows, exhibitions and other activities to educate and promote car-lite options. Moreover, the government conducts workshops and seminars to encourage and educate people on the benefits of walking. These events aim to promote the importance of physical activity, such as walking, for health and well-being. The events also highlight the environmental benefits of walking, such as reducing carbon emissions, and the economic benefits, such as saving money on transportation costs.

Overall, the Singapore government's commendable efforts in promoting walking deserve recognition, considering the challenge of changing public perception and encouraging individuals to view walking as a viable mode of transportation and incorporate it into their daily lives.

[16] See SG Press Centre (2012).

Cycling in Singapore

Cycling is often regarded as a fun and enjoyable activity. As compared to walking, cyclists can travel further and faster, though the length of distance is nonetheless limited. Cycling is also a preferred mode of transport among environmentalists as it helps to reduce carbon footprint.

The demand for cycling is influenced by a string of factors. These include the users' perception and the built environment, safety matters and convenient provisions for cycling.[17] Rojas Lopez and Wong Yiik-Diew conducted an extensive study with adults and elderly to gather the perceived opportunities, concerns, advantages and disadvantages to encourage walking and cycling in Singapore.[18] Users were contented with the walking infrastructure, though they conveyed that more could be done to upgrade the cycling infrastructure such as widening the cycling pathways and increasing the cycling bay facilities. Notably, more positive attitude towards cycling was recorded as compared to walking. Users opined that flexible working hours allow them to do more cycling. Cyclists can engage in both active modes from the first to the last mile of their trips, for example, from mass rapid or light rail stations to home.[19]

In recent years, bicycle-sharing programmes have been established in Singapore as means of providing residents and visitors with a convenient

[17] See Alfred and Goodman (2020); Easton & Ferrari (2015); Eldeeb et al. (2021); Gehrke and Wang (2020); Guzman et al. (2020); Tao et al. (2020). In 1885, British inventors John Kemp Starley and William Sutton offered the Rover bicycles with two equally sized wheels, a chain, direct steering and a stell frame. The design improved in 1888 with John Dunlop's pneumatic tires. Interestingly, unlike cars, which has changed dramatically over the years, bicycles design has remained remarkably consistent and conservative.

[18] Lopez and Wong (2017).

[19] Public transportation in Singapore is extensive and generally affordable. It encourages people to ride their first and last mile trips on motorised vehicles which can save them time to reach their next destination. Parents who desire to walk or cycle more can park their cars at home after dropping off their young children at schools and proceed to work by public transportation. The bus service enhancement programme was carefully designed to increase the number of buses on the road. By expanding the MRT network to approximately 360 kilometres by 2030, Singapore aims to enhance accessibility and connectivity for its residents. This ambitious plan is set to bring significant benefits, with the goal of ensuring that 8 out of 10 households will be within a convenient 10-minute walk of a nearby train station.

and eco-friendly mode of transportation.[20] These programmes allow individuals to easily rent a bike for a short period of time, whether it be for leisure or commuting, with the goal of reducing traffic congestion and promoting a more sustainable mode of transportation. Technology advancements contribute to an explosion of IT-based systems that make bike sharing possible. All the users need is a smartphone connected to an app. After locating the bike, the user unlocks it by scanning the QR. Bike operators put a time limit to bike use, track the location of the bikes using GPS, collect payments online and ensure the anonymity of users and payment. Bike sharing however posed several challenges. First, there was a lack of dedicated space for cyclists, many of whom had to share off-road pavements with pedestrians and on-road lanes with motorists. Second, the bike-sharing services were dockless, which means that the bikes could be parked in any place. But not all users would obediently park their bikes in dedicated bike-parking lots. Many left their bikes along the road, inducing conflicts in urban management.

Indiscriminate parking by users and abuse of the bicycles led to more stringent regulations from the Singapore government. By October 2017, five bicycle-sharing companies signed a memorandum of understanding (MOU) with Singapore's LTA, NParks and town councils to remove indiscriminately parked bicycles within half a day, implement geo-fencing technologies by the end of 2017 (the solution requires users to scan a Quick Response or QR code at each parking space to ensure proper parking) and educate users on the proper use of their bicycles. In March 2018, the Singapore parliament passed the Parking Places (Amendment) Bill, which required the operators, among other things, to renew their licenses every six months as part of the government's effort to review the fleet size. Ofo abruptly exited the Singapore market in 2019 in the wake of new regulations implemented by the LTA while Mobike, another bike operator, surrendered its license a year later, leaving homegrown SG Bike as the sole provider with full licenses.

All is not lost as Singapore continues to invest in cycling infrastructure. Indeed, back in 2013, the Land Use Plan called for road cycling path networks and bicycle parking facilities in all HDB towns as part of the National Cycling Plan. There was also the Round Island Route

[20] In 2016, new set-up companies like Ofo and Mobike offered bike-sharing services in Singapore.

initiative—a 150-kilometre route to provide an accessible and uninterrupted leisure corridor around the island, and the Rail Corridor—along the old Malayan Railway track to offer inviting routes for joggers and cyclists and improve access to recreational and heritage sites—to improve accessibility and connectivity. In January 2022, Deputy Prime Minister Heng Swee Keat commemorated the opening of the first half of the Round Island Route, a part of the government's effort to transform Singapore into a City of Nature under the Singapore Green Plan 2030. The 150-kilometre route, covering the east coast of Singapore and ending at Berlayer Creek near Labrador Nature Reserve, provides pedestrians and cyclists a "seamless ride or ramble from the northeast to the south of Singapore", and is Singapore's longest recreational route to date. The remaining 75 kilometres of the route is expected to complete by 2035.

Cycling within towns to connect to schools, facilities and amenities in an efficient and reliable way are supported by dedicated off-road cycling paths in HDB towns such as Tampines. The longer term is to connect off-road cycling paths to a wide section of the island, tapping on the extensive PCNs to establish a large and safe cycling network using under-utilised land such as drainage reserves, foreshores and road reserves, to turn into green corridors for recreation and use by pedestrians and cyclists. The creation of parks and green open spaces, as part of the greening programme, caters to residents' desire for open spaces for play and leisure purposes. The first stretched was opened in 1995—the Kallang Park Connector, linking Bishan-Ang Mo Kio Park and Kallang Riverside Park. To optimise land use, the network has been designed to run on covered drains, under MRT viaducts and on adjoining public housing land.

LTA's Islandwide Cycling Network programme was unveiled in 2020 and aims to expand Singapore cycling path network to span about 1,300 kilometres by 2030. The plan includes the construction of off-road cycling paths, which are typically between 2 metres and 2.5 metres in width, with coloured pavement (usually red, blue, or grey). Intra-town cycling networks are implemented in several townships in Singapore to provide a safe and convenient way for cyclists to commute relatively long distances with the current length of cycling path islandwide exceeding 450 kilometres. To support the initiative, dedicated off-road cycling paths are constructed to segregate the cyclists from pedestrians and motorists. Broadly speaking, two types of cycling facilities are available in Singapore.

- Dedicated on-road cycling lane—cyclists are provided with their own space on the road, which they share with motorists. The two groups of users are separated by coloured demarcation lines. Typically, the dedicated on-road cycling lane is between 1.5 metres and 2 metres in width. Exclusive lane for cyclists was first introduced in West Camp Road in October 2022. Marked out by solid blue lines, the 4.6 kilometres stretch on both directions of the road grant exclusive rights to cycling during fixed hours—5 a.m. to 11 a.m. on Sundays. The trial ended in April 2023 before other areas are replicated for cyclists to use recreationally and to go to work.
- Shared off-road footway—cyclists share the footway space with the pedestrians. Prior to the enactment of the Active Mobility Act 2017, cyclists were forbidden by law to share off-road footways with pedestrians and would therefore have to travel on the road and share the road lanes with motorists. The width of the road lanes varies. The tendency for motorised-related traffic accidents increased on narrower roads. Cyclists are particularly vulnerable due to huge differences in kinetic energy and safety protection between cyclists and motorists, resulting in severe casualties to cyclists. Between 2013 and 2017, the number of cycling casualties increased from 451 to 604.[21] The Active Mobility Act 2017 was passed by Parliament on 10 January 2017 to enable pedestrians and cyclists to share footpaths safely. The Act regulates the use of public paths for walking, cycling or other similar purposes. The Act stipulates that all riders of Active Mobility devices such as bicycles, Power-Assisted Bicycles (PABs), Personal Mobility Devices (PMDs) and Personal Mobility Aids (PMAs) are required to adhere to rules under the Active Mobility Act, which include non-usage of mobile communication devices while riding, dismounting and pushing the device when the rider sees a "No Riding" sign (with the exception of PMA users) and switching on the front white and rear red lights when riding in the dark.

To bolster the cycling infrastructure in Singapore, a progressive approach is being taken to construct cycling paths, cycling bridges, bicycle

[21] Extracted from Singapore Yearbook of Statistics 2018 published by the Department of Statistics Singapore. See also Elvik (2010).

parking spaces and more in different towns and parts of the city with the overall aim of expanding Singapore's cycling network to an impressive 1,300 kilometres by 2030. Singapore's first walking and cycling town, Ang Mo Kio is home to a 4-kilometre cycling network linking its dwellers to Ang Mo Kio bus interchange and MRT station. Improved access to Ang Mo Kio Hub and the swimming complex in all had made it more seamless and safer for people to walk and cycle the first and last mile for their daily travelling. An extension of 20-kilometre cycling paths is underway, which will provide the longest route for any township. Ang Mo Kio had installed several innovative solutions that enable cycling and walking in its built-up environment. New junction design, planning typologies, innovative elevated infrastructure and harmonised signage for cycling and walking have been tested. Pedestrian and cyclists can enjoy special features from easy-to-read maps on signboards, clearly marked pedestrian priority zones, red cycling paths, bicycle wheeling ramps along staircases and space conversion of the MRT viaduct into sheltered cycling paths amid greenery applying horticultural technology in collaboration with the NParks. Future plans include designing a direct cycling route from the city to other towns in the North–South Corridor such as Sembawang and Yishun as well as other townships such as Geylang and Queenstown.

On 28 November 2022, LTA unveiled the 6-kilometre cycling path in the CBD, connecting workplaces, heritage sites and amenities. "Here in the Central Business District and heritage district", said Transport Minister S. Iswaran at the launch, "there is great potential to make walking and cycling the preferred way to get around".[22] Fifty-five kilometres of cycling path is expected in Geylang, Hougang, Kaki Bukit, Marine Parade, Potong Pasir, Sengkang and Serangoon by the end of 2024, adding to the 30-kilometre path that has been completed in Ang Mo Kio, Bishan, the Central Business District, Taman Jurong, Tampines and Toa Payoh.[23] Properly demarcated, the paths are connected to amenities, such as transport nodes and schools, to encourage cycling around towns and supported by infrastructure such as bicycle parking facilities, bicycle crossings, wheeling ramps and map board. Constructing the paths

[22] Ng (2022, November 28).

[23] Kok (2023, January 13).

requires repurposing existing road space such as drains and bus stops to allow for the new paths.

Bencoolen Street has undergone a significant transformation whereby two of its four vehicular lanes have cycling lanes and walkways broadened alongside the street. It was an opportune time to pedestrianise the street which was condoned off from public use during the Downtown Line's construction of the Bencoonlen station. There are dedicated bus lanes, custom-built cycling paths linking to the other parts of the city or the rest of Singapore, wider footpaths and benches for pedestrians to rest under shady foliage as well as lined with more than a hundred bicycle parking spaces. Similar plans have been emulated in other areas including Armenian Street, Coleman Street and Waterloo Street.

One of the popular cycling routes is the 10 kilometres Tanah Merah Coast Road, which offers one-way cycling infrastructure that bears new markings to increase cyclists' safety alongside the stretch. Alternatively, causal cyclists may opt to use the PCNs running shoulder-to-shoulder with the Tanah Merah Coast Road especially those unfamiliar with the road traffic. In October 2021, a new 5.6 kilometres cycling path was opened in Taman Jurong, connecting Lakeside and Chinese Garden MRT stations. In December 2021, the government announced the plan to reclaim parts of the roads (totalling 50 roads), including West Coast Highway, Commonwealth Avenue West, Hume Avenue, Toh Tuck Avenue and Boon Lay Way, to make space for the construction of 120 kilometres of cycling paths in Singapore. The cycling paths are separated from the roads for motorised vehicles and built next to footpaths thereby significantly improving the safety of users. Expansion plans for the existing cycling path networks in Tampines and Ang Mo Kio are expected to be completed by 2024. Dwellers in Tampines town are connected to Changi Business Park and neighbourhood towns with these additional routes. Other enhancements include connecting Geylang to the City via the Geylang Park Connector and linking Queenstown to the City by passing through the Alexandra Park Connector.

The efforts have yielded positive outcomes. The proportion of Singaporeans who aspire to own a car has declined whereas people are cycling more often than before. In a survey commission by the Straits Times and conducted by Nexus Link in April 2022, 50.4% of the respondents said they aspire to own a car within the next few years, down from 65.5%

in 2016.[24] Over 5,000 respondents between 18 and 35 years old took part in the survey; 55% of them earned between S$2,000 and S$5,000 a month and 74% of them had polytechnic education qualification and above. The main reason for giving up owning a car was proximity of their homes to either a train station or a bus stop; about 90% of the respondents lived within 10 minutes' walk of a bus stop and 44.3% lived within 10 minutes' walk of a train station. The same study found more people cycling. 50.5% of the respondents said that they cycle more than five times a week, up from 27.4% in 2016. The survey results indicate a positive impact of the government's initiatives aimed at promoting and encouraging walking and cycling.

In order to truly appreciate the transformation, it is crucial to consider the historical conditions of urban space in Singapore. The street patterns in Singapore after 1960 were not designed to promote walking and cycling. Streets are often inhospitable to pedestrians. Block sizes are often too large and within proximity to one another to allow for wide sidewalks or other land-use patterns to enable pedestrians and cyclists to navigate freely. By the 1970s, as the population became more affluent, cars and motorcycles gained popularity. The use of bicycles to travel from point to point declined significantly. Cycling paths were removed to widen the roads to serve the needs of motorists. Like in many countries, pedestrians were kept to pavements and the term jaywalking was negatively connotated to refer to someone who does not know how to walk in the city.

Despite the progress made, challenges persist, with road safety remaining a top priority. To a large extent, the pathways in urban spaces and neighbourhoods are shared by cyclists, pedestrians and personal mobility device (PMD) riders who are on their hoverboards and kick scooters. The principal danger is the varying speed of PMD riders and cyclists, often faster, than the walking pedestrians, who could miss looking around for other users. Waiting areas of road crossings and bus stops located along the pathways become a pathway hazard especially during peak hours, which may cause other pedestrians to knock into other pedestrians or road accidents when pedestrians are eager to ease from the

[24] Lee (2022, June 18).

rushing crowd and to their destinations. Elderly and kids are more vulnerable users and more likely involve in a pathway or road accident due to pathways being too cramped or when walking along heavy traffic roads.

Cyclists riding on pathways are also concerned as they may be unable to act quickly if other users (e.g., kids and wheelchair-bound users) suddenly appear in their direct cycling path. Often, cyclists are looked upon as *culprits* and the ones injured as *victimized* in the accidents. In these events, cyclists could be subjected to public scorn for not being careful enough which can create phobia of the activity in extreme cases. Safety concerns are exacerbated by cyclists not heeding safety measures such as wearing proper safety gears like helmets and bright-coloured clothes as well as possessing a well-lit bicycle. This can make it difficult for other users to identify them, henceforth increasing possible accidents and conflicts. Notwithstanding the fact that motorists pay to use the road, Chua Mui Hoong feels that motorists have the right to roads more than pedestrians and cyclists, citing the drivers' apparent entitlement to drive comfortably on a three-lane road while cyclists and pedestrians are "confined to a narrow strip of pavement maybe 1.5 m wide".[25] Cyclists can of course ride on roads, she wrote, "but most do not for safety reasons". It seems that in Singapore, "if you own and drive a car, you are among the 20% of the population who do so. You are privileged road user, with roads designed to your advantage; and other road users confined to the margins—pedestrians on pavements, cyclists on the left land, near the kerb".

Another challenge is that, while more people are cycling, they do so for leisure purposes rather than cycling to work. The key reasons cited were the long distance, hot weather and safety concerns. Workplaces that offer shower facilities are still limited which is not very encouraging for people considering cycling to work. Turning up for work sweaty in work attire is not an acceptable grooming standard in general. Some may find it uncomfortable despite bringing another set of work wear to change without having a shower.

Notable strides have been taken to address these issues. LTA has taken a proactive approach in partnering with developers and building owners in business parks, light and general industrial areas and retail developers

[25] Chua (2022, May 27).

to promote active commuting and improve the overall commuting experience. As part of this collaboration, they have facilitated the provision of end-of-trip facilities such as lockers, changing rooms and shower rooms. To incentivise developers, the Active Commute Grant offers reimbursement of up to 80% of the construction cost for these facilities, with a maximum grant of up to S$80,000 per development. The creation of more road spaces for cycling, linking the outskirts of the city to the city area—the North–South Corridor, for example—may also help to address some of the concerns. In terms of safety, the introduction of Ridenjoy, Singapore's innovative active mobility lifestyle app, provides cyclists with a convenient and secure means of travelling using dedicated park connectors and cycling paths. Instead of relying on Google Maps or Apple Maps, which may direct cyclists to busy and dangerous roads, the app utilises Singapore Land Authority's OneMap API to recommend cycling using only park connectors. Launched in December 2021 by Singaporean Kelvin Phang, Ridenjoy considers safety as the top priority for any form of cycling. Phang wrote "After trying many apps, I realize I can cycle to downtown safely using only park connectors and cycling paths. Thanks to Sport Singapore's (SportSG) 'Blended' initiative, we have developed an app that enables cyclists to safely navigate around Singapore without using public roads and to truly enjoy the Park Connector Network (PCN) and Islandwide Cycling Network (ICN)."[26]

In sum, Singapore has made substantial investments in walkability infrastructure, reflecting a growing emphasis on environmental consciousness and congestion mitigation. This chapter provides clear evidence of this. The built environment is recognised by urban planners in Singapore as a factor influencing travel behaviour; hence trees and greenery are often seen as desirable for urban walkability. To encourage walking, the LTA has built covered linkways to provide a sheltered walkway network between transport nodes and various amenities. They are built based on the following guidelines, first, between MRT stations and trip-generating hubs such as schools, healthcare facilities, public amenities and commercial and residential developments within a 400-metre radius of the stations; and second, between bus interchanges and LRT stations and developments within a 200 metres radius. Covered linkways with

[26] Quoted in https://www.ridenjoy.sg/make-the-most-of-the-season-by-following-these-simple-guidelines. Accessed on 18 July 2022.

LED lightings are complemented with rest areas at recommended intervals of 70 metres to 100 metres. Such rest areas comprise a bench and wheelchair parking space. New footpaths have increased from 1.5 metres to 1.8 metres, which is the adequate width for two typical wheelchairs to pass each other, and more than 1.8 metres for walkways with a higher volume of pedestrians. Furthermore, LTA and URA require development work for selected residential, commercial, industrial, educational, medical, hotel and recreational developments to submit street work proposals to ensure that walkability is taken into consideration in the planning. Among other things, tactile tiles and kerb ramps must be provided at areas near crossings to remind pedestrians, especially pedestrians with vision impairment, to be vigilant while crossing the road. Developers are also to ensure that sufficient roadside verge is provided for tree planting to sustain the pervasive sense of greenery and offer a more pleasant environment for pedestrians and cyclists.

Giving priority to cyclists and pedestrians is a significant change in Singapore as part of a large effort to promote sustainable development and reduce traffic congestion. By encouraging the use of active modes of transportation such as walking and cycling, the government hopes to decrease the number of cars on the road, improve air quality and create a more liveable city for residents, in addition to promoting physical activity and improve public health. Overall, this change in Singapore is a response to the growing global trend towards sustainable transportation urban planning. To further grasp the essence of walkability in Singapore, the upcoming three chapters offer an immersive exploration of the city-state's vibrant cultural and natural landscapes. Within these chapters, you will discover a tapestry of heritage sites, magnificent architectural wonders, captivating islands, picturesque trails and bustling parks, all of which contribute to the remarkable walkability experience.

References

Alfred, R., & Goodman, A. (2020). Low traffic neighbourhoods, car use, and active travel: Evidence from the people and places survey of outer London active travel interventions. *Transport Findings*. https://doi.org/10.32866/001c.17128. Accessed on 14 September 2021.

Cheam, J. (2009a, December 3). Singapore to pledge 16% cut in gas emissions. *The Straits Times*.

Cheam, J. (2009b, December 3). Singapore wants to help bring about a deal. *The Straits Times*.

Chua, M. H. (2022, May 27). How 'jaywalkers' lost the battle for road space to cars. *The Straits Times*.

Easton, S., & Ferrari, E. (2015). Children's travel to school—The interaction of individual, neighbourhood and school factors. *Transport Policy*, *44*, 9–18. https://doi.org/10.1016/j.tranpol.2015.05.023

Eldeeb, G., Mohamed, M., & Páez, A. (2021). Built for active travel? Investigating the contextual effects of the built environment on transportation mode choice. *Journal of Transport Geography*, *96*. https://doi.org/10.1016/j.jtrangeo.2021.103158

Elvik, R. (2010). Why some road safety problems are more difficult to solve than others. *Accident Analysis and Prevention*, *42*(4), 1089–1096.

Gehrke, S. R., & Wang, L. (2020). Operationalizing the neighborhood effects of the built environment on travel behavior. *Journal of Transport Geography*, *82*. https://doi.org/10.1016/j.jtrangeo.2019.102561

Government of Singapore. (2022, March 5). *Factsheet: 50 silver zones by 2023 to enhance road safety for senior pedestrians*. https://www.lta.gov.sg/content/ltagov/en/newsroom/2020/3/news-releases/50-silver-zones-by-2023-to-enhance-road-safety-for-senior-pedest.html#:~:text=Together%20with%20the%2017%20Silver,will%20be%20completed%20by%202023. Accessed on 11 May 2022.

Gunasingham, A. (2009, October 30). Singapore rejects emission cuts. *The Straits Times*.

Guzman, L. A., Peña, J., & Carrasco, J. A. (2020). Assessing the role of the built environment and sociodemographic characteristics on walking travel distances in Bogotá. *Journal of Transport Geography*, *88*. https://doi.org/10.1016/j.jtrangeo.2020.102844

Ho, A. (2009, October 30). Reasons for Singapore to be cool on global warming. *The Straits Times*.

Koh, P. P., & Wong, Y. D. (2013). Comparing pedestrians' needs and behaviours in different land use environments. *Journal of Transport Geography, 26*, 43–50.

Kok, Y. F. (2023, January 13). New cycling paths in 7 towns to be ready from 2024. *The Straits Times*.

Lee, M. T. (2022, June 18). Car aspirations wane: Few are willing to commute on bicycle. *The Straits Times*.

Lopez, M. C. R., & Wong, Y. D. (2017). Attitudes towards active mobility in Singapore: A qualitative study. *Case Studies on Transport Policy,* 5(4), 662–670.

Ng, W. K. (2022, November 28). New CBD cycling paths link workplaces, MRT stations. *The Straits Times*.

Ng, W. S., & Mendelsohn, R. (2005). The impact of sea level rise on Singapore. *Environment and Development Economics, 10*(2), 201–225.

Nur Dianah Suhaimi (2009, October 31). Emissions cut if others do their part as well. *The Straits Times.*

Olszewski, P., & Wibowo, S. S. (2005). Using equivalent walking distance to assess pedestrian accessibility to transit stations in Singapore. *Transportation Research Record: Journal of the Transportation Research Board, 1927*(1), 38–45.

SG Press Centre. (2012, July 1). *Speech by Mr. Lee Hsien Loong*. World Cities Summit, Singapore.

Tao, T., Wu, X., Cao, J., Fan, Y., Das, K., & Ramaswami, A. (2020). Exploring the nonlinear relationship between the built environment and active travel in the twin cities. *Journal of Planning Education and Research.* https://doi.org/10.1177%2F0739456X20915765

CHAPTER 4

Contemporary Heritage-Built Attractions

During his speech in Parliament on 5 November 1970, Edmund W. Barker, the National Development Minister at the time, expressed concerns about the negative impact of neglect and rapid urban renewal on Singapore's historical monuments. He emphasised the need to protect these monuments, stating, "In our eagerness for urban renewal, we must not neglect our historic treasures, which may one day be lost to us if we are not careful".[1] Fast forward to the present, the protection of Singapore's built heritage has become an integral part of the national urban planning strategy, ensuring the preservation of what remains of the historical monuments.

This chapter will discuss Singapore's past, preservation and future preservation efforts, with a deliberate focus on walkability and Singapore's existing cultural heritage walking trails in its historic enclaves and other areas. The challenges in maintaining Singapore's cultural heritage-built attractions and concrete actions supporting specific United Nations Sustainable Development Goals (SDGs) are also discussed.

[1] SG101 (2022).

Melissa Liow Li Sa and Sam Choon-Yin, *Sustainable Urban Development in Singapore*, Sustainable Development Goals Series, https://doi.org/10.1007/978-981-99-5451-3_4

Past, Present and Future Preservation Efforts

The observable preservation efforts by Singapore started with the formation of the Committee for the Preservation of Historic Sites and Antiquities in 1950, to undertake the role in preserving sites and secular/non-secular buildings with historical significance. The Preservation of Monuments Act was legislated in 1971 to allow for the establishment of the Preservation of Monuments Board (PMB). The board is designated to preserve monuments of archaeological, historic, architectural, traditional or artistic interest for the benefit of the nation.[2]

URA's Conservation Master Plan was introduced in 1986, offering Singapore the first blueprint to conserve the nation's-built heritage. URA carried out numerous projects in restoring rows of colonial shop–house-shop–houses at Cuppage Road, Emerald Hill, Murray Street and Tanglin Road. This consequently developed into a comprehensive master plan to include six areas namely Chinatown, Emerald Hill, Heritage Link (a cultural and civic strip connecting Bras Basah Road, Empress Place, Fort Canning Park), Kampong Glam, Little India and Singapore River earmarked as historical sites for URA's conservation.[3] Today, Singapore persists in conserving its historic districts and protecting its built heritage owing to the paired efforts of URA and Preservation of Sites and Monuments (PSM), the successor of PMB.[4] There is hope for heritage areas and historically significant buildings that have outlived the ravages of time. The Singapore government endeavours to preserve its physical landscape and the accompanying special characteristics to give a sense of place for community gatherings, interaction and for younger generations to grow up.

Table 4.1 provides a summary of the past, present and future preservation efforts of the heritage-built attractions.

The significance of a nation's culture and heritage identity for attaining sustainable development has been recognised as *the fourth pillar of sustainability* (that comes after economic, social and environmental pillars of sustainability) under the United Cities and Local Governments

[2] Government of Singapore (2023a).

[3] Government of Singapore (2023b).

[4] National Library Board, Singapore (2019, July 31).

Table 4.1 Heritage preservation in Singapore

Time eras in Singapore	*Preservation efforts*	*Key stakeholders*
1950s: Early Colonial Efforts	1951: Preservation of the last ruler of the fourteenth-century Singapura, Sultan Iskandar Shah's tomb and nineteenth-century Christian cemetery on Fort Canning Hill, which later was converted into a public park The wear-and-tear of the tombs due to years of neglect were either recovered and implanted into the park walls or those intact are maintained for their historical value	1951: Michael W. F. Tweedie, who was then Director of the Raffles Museum
	1954: The committee team had been requested to put up a list of historic sites (30 sites in total, majority from the nineteenth century) in Singapore with plaques in English: but if Chinese or Malay origins, an accompanying English inscription will be placed too) at these sites to mark their historical significance. Examples of the sites include Saint Andrew's Cathedral, Cathedral of the Good Shepherd, Coleman House (was demolished to make way for the present Peninsula Shopping Centre in 1965), Elgin Bridge, *Masjid* (Mosque in Malay) Hajjah Fatimah, Old Parliament House, Sri Mariamman Temple, Thian Hock Keng temple and Victoria Theatre	1954: T.H.H. Hancock and Carl Alexander Gibson-Hill—senior architect of the Public Works Department and curator of zoology at the Raffles Museum. Both are from the Committee for the Preservation of Historic Sites and Antiquities. Singapore Improvement Trust (SIT) in consultation with the members from the Committee for the Preservation of Historic Sites and Antiquities and Friends of Singapore founded in 1937 by leading public figures (such as Song Ong Siang, a key member from the Straits Chinese community and a famous lawyer, Roland St John Braddell). Friends of Singapore

(continued)

Table 4.1 (continued)

Time eras in Singapore	*Preservation efforts*	*Key stakeholders*
	The committee's role is to restore the identified historic buildings and preserve them for future generations. As many are privately owned, lots of government funds to acquire these historic buildings/structure would be needed. Singapore Improvement Trust (SIT) was tasked to update the list of ancient monuments with architectural/historic interest for consideration in preservation efforts under the 1958 Master Plan and 1959 Planning Ordinance. New additions to the list include the Indian Cemetery in Geylang, Kampong Radin Mas Cemetery, 3 Oxley Rise (or better known as Killiney House) and Outram Gaol Friends of Singapore have made some achievements though it was difficult in the beginning. The team had to launch several public campaigns to preserve the 1942 Pasir Panjang battle site when the Malay regiment put up a fight against the Japanese army (a commemorative park was created); restoring the Coleman House built in 1929 and converted the building into a *home of the arts*, demarcate the Ulu Pandan and Bukit Timah nature sites as proper nature parks and to keep the Killiney house (an 1842 planters home that symbolises Straits Settlement, which was constructed by then surgeon-general, Thomas Oxley) which was almost erased from the ancient monument list in the 1958 Master Plan heritage site list	

Time eras in Singapore	*Preservation efforts*	*Key stakeholders*
	1959: Singapore's People Action Party (PAP) became the ruling party but there were more pressing needs (250 thousand people are living in overcrowded slums, one-third in squatters, mostly in dilapidated state in the city centre, so rehousing matters) became the top priority. Preserving built heritage was, unfortunately, given the least attention	1959: Singapore's People Action Party (PAP)
1960s—Urban Renewal and Demolition; Harbour Rehabilitation Efforts	1960: A hard-line public housing programme was initiated that saw new housing estates constructed but located outside the city centre. Redevelopment of the main area is one of the key tasks under the urban renewal programme Clearing of slums and people resettlement away from the city centre, and demolition of old buildings of 1,500 acres of the old city was to make way for constructing new buildings and develop an integrated modern city. Land value was maximised in this event, but the matter of preserving historical buildings and sites was not seen as important	Housing & Development Board (HDB) replaced SIT Urban Renewal Department (URD)

(continued)

Table 4.1 (continued)

Time eras in Singapore	*Preservation efforts*	*Key stakeholders*
	1962 to 1963: Government was aware of the need to preserve the city's historic sites in spite of the random building/structure demolition. The United Nations team explained that urban renewal need not necessarily demolish old buildings, but to institute—*conservation, rehabilitation and rebuilding* values	United Nations town planning advisor, Erik Lorange was engaged in 1962 and a second United Nations team was despatched in 1963. The recommendations were backed by Singaporean architects like Tay Kheng Soon and William Lim including the urban planning think tank of the duo who founded SPUR—Singapore Planning and Urban Research Group
	The Singapore Planning and Urban Research Group (SPUR) cited in its 1967 periodical, "redevelopment is necessary as part of the evolution of any City" but the magnitude of redevelopment to be kept at a minimum while upholding the three processes—*conservation, rehabilitation and rebuilding* as recommended by the United Nations team[5]	
	1964: Outram Gaol, Coleman House, Killiney House and Raffles Institution were either razed or pulled down to give space for building housing flats or commercial plans under the 1964 Precinct South 1	Housing & Development Board (HDB)
	Since the official launch of the urban renewal programme in 1966, in less than a decade, HDB has constructed 130 thousand flats in new housing estates to relocate 40% of the population while redeveloped close to 300 acres of the city centre	

[5] SG101 (2022).

Time eras in Singapore	*Preservation efforts*	*Key stakeholders*
	1967: Then Prime Minister Lee Kuan Yew wrote with praise to URD for preserving buildings and what little there is of historic interest such as Hajjah Fatimah Mosque in the redevelopment stage of Precinct North 2B	Urban Renewal Department (URD)
1970s—Establishment of Preservation of Monuments Board (PMB), Historic Buildings to Historic Districts	1971: The Preservation of Monuments Act was ratified, and the Preservation of Monuments Board (PMB) was established. The primary function of the board is to identify, preserve and maintain historic landmarks linked to archaeological, architectural, artistic, cultural or historic interest that represent links to Singapore's past. Other roles include advocacy, document information on gazetted monuments and launch promotion to preserve and restore monuments composed of civic, commercial, cultural and religious building structures 28 June 1973: The monuments that were gazetted by PMB include Armenian Church, Cathedral of the Good Shepherd, Hajjah Fatimah Mosque, Old Thong Chai Medical Institution, Sri Mariamman Temple, St Andrew's Cathedral, Telok Ayer Market (Lau Pa Sat) and Thian Hock Keng Temple	Preservation of Monuments Board (PMB)

(continued)

Table 4.1 (continued)

Time eras in Singapore	*Preservation efforts*	*Key stakeholders*
	1974: URA expanded the preservation of Singapore's built heritage to another level by not only retaining architectural, cultural and historical buildings but also to rehabilitation and conserving the entire district's character or traditional area settings. This integrated approach was first undertaken to convert 14 Deco colonial shop-houses into restaurants on Murray Street, 17 Malacca-style terrace houses for commercial use on Cuppage Road, six colonial shop-houses as a pedestrian Peranakan mall on Emerald Hill Road and 9 Tudor-style former government quarters into a shopping mall and offices on Tanglin Road	Urban Redevelopment Authority (URA) superseding URD in 1974
	1977: The World Heritage Committee organised its first session. The committee comprises 21 States Parties elected by its General Assembly. The primary role of committee is to define and budget funds dedicated to carrying out the conservation of inscribed properties from the World Heritage List or obliterate properties from the List of World Heritage in Danger	The World Heritage Committee

Time eras in Singapore	*Preservation efforts*	*Key stakeholders*
1980s—Conservation Blueprint	1982: URA conducted a review of Singapore's city centre urban design structure and developed a conservation blueprint including conservation guidelines to conserve the historical character of the area 1986: Six historic areas—Chinatown, Emerald Hill, Heritage Link (comprises of Bras Basah Road, Empress Place and Fort Canning Park), Little India, Kampong Glam, Singapore River were designated for conservation covering 4% of the main area. Preserving these sites was deemed critical to retain Singapore's city landscape and its unique identity 1987: URA restored 32 old shop-houses in Tanjong Pagar and expanded the conservation efforts to revive another 220 state-owned shop-houses in the surroundings to give a new lease of life to *long-hidden beauties* (i.e., the old shop-houses) located across several lanes of a whole conserved area 1989: A master plan was underway under URA's conservation blueprint to gazette as the pioneer nation's historic districts (Boat Quay, Clarke Quay, Cairnhill, Chinatown, Emerald Hill, Little India and Kampong Glam) The Planning Act was revised to empower URA as the national conservation authority, the main government body to classify areas for conservation and the approving authority for developers wanting to perform restoration works in conserved areas	Urban Redevelopment Authority (URA) Urban Redevelopment Authority (URA)

(continued)

Table 4.1 (continued)

Time eras in Singapore	*Preservation efforts*	*Key stakeholders*
1990s—Milestones and the Way Forward	1973 to 2018: Gazetted national monuments grew from 8 to 72. Heritage markers were placed by PSM at places of historical significance with momentous events and well-known personalities linked to each of the places	Preservation of Monuments Board (PMB) was a statutory board under the Ministry of National Development before transferring under the Ministry of Communications and Information with effective from 1997 and then combined with National Heritage Board in 2009, and thereafter redesignated as Preservation of Sites and Monuments (PSM) in 2013
	1 August 1993: National Heritage Board (NHB) was established and is the custodian of Singapore's heritage, where it is responsible to promote, commemorate and preserve the both intangible and tangible cultural heritage of Singapore. The board takes charge of telling the Great Singapore story, conveying the Singapore spirit and the Singaporean experiences in past nation-building efforts, and educating the public to invoke a better cultural understanding of the shared heritage of multi-ethnic communities. It manages and implements policies of Singapore's heritage institutions, national museums, heritage sites and arts and cultural belts such as the Bras Basah and Bugis precincts. Other roles include preserving Singapore's heritage trails, national collection and monuments and documenting the nation's communal beliefs, cultural legacies, lifestyles and traditions through community projects, exhibitions and festivals that make up the Singaporean society	National Heritage Board (NHB) is a constitutional board parked under Ministry of Culture, Community and Youth

Time eras in Singapore	*Preservation efforts*	*Key stakeholders*
	1995: URA presented the inaugural Architectural Heritage Awards to architects, contractors, developers and engineers who have exhibited excellent conservation and restoration efforts in Singapore. Since then, more than 100 buildings have been conferred with the award including the 11 Kim Yam Road, 77 Emerald Hill Road, 149 Neil Road, 161 Lavender Road, Armenian Church, Cathedral of the Good Shepherd, Chijmes, River House in Clarke Quay and Sultan Mosque In 2015, the Singapore Botanic Gardens was awarded the status of UNESCO World Heritage Site (WHS) 2017: The Architectural Heritage Season was launched. The festival marks Singapore's progress in preserving built heritage and carefully restored monuments. Community partners, such as students, professionals and volunteers act as docents to share the technical restoration details with the public in the form of guided tours As of 2018: URA continued its efforts of updating conservation guidelines and demarcating new conserved areas. Over 1,000 locations and 7,000 buildings have been conserved, primarily retaining the spatial and architectural characteristics of buildings while enabling adaptive reuse. Careful repair, maximum retention and sensitive restoration are 3 URA's core conservation principles which are taken into consideration for these conservation projects	Urban Redevelopment Authority (URA) UNESCO World Heritage representatives Urban Redevelopment Authority (URA) UNESCO Intangible Cultural Heritage (1992–2022)

(continued)

Table 4.1 (continued)

Time eras in Singapore	*Preservation efforts*	*Key stakeholders*
	16 December 2020: Singapore's hawker culture in Singapore, culinary practices and community dining in a multicultural urban context were inscribed in 2020 on the UNESCO Representative List of the Intangible Cultural Heritage of Humanity. Hawker centres (such as *Lau Pa Sat)* acts as all-day community dining rooms for people from different walks of life to meet, dine and enjoy recreational activities such as art-jamming, busking and chess-playing. The oldest hawkers started their small business in the 1960s, often specialising in a specific dish, gradually refining the recipes from the older generation to the younger generation (either their apprentices or family members) 28 September 2022: The first Singaporean hawker centre of 14,000 square feet in the Manhattan borough of New York City, is modelled after the nation's famous hawker markets, offering Chinese, Indian, Malay, Peranakan, Southeast Asian and British-influenced hawker food fare/favourites from 11 renowned Singapore hawker vendors from a total 17 stalls	

Source Domrongchai (2022, September 28); National Library Board, Singapore. (2019, July 31); SG101 (2022). UNESCO Intangible Cultural Heritage (1992–2022)

(UCLG) policy document.[6] The Sustainable Development Goals (SDGs) of Agenda 2030 have similarly paid attention to the preservation of heritage, particularly Sustainable Development Goals 8 and 11.4 aims to "strengthen efforts to protect and safeguard the world's cultural and natural heritage".[7] Singapore supports the 2030 agenda. Singapore partners United Nations' agencies such as UNICEF and UN-Habitat to run capacity-building courses in water-related issues and sustainable urbanisation. Singapore adopts the Whole-of-Government (WOG) approach, which entails the sharing of information among public agencies to incorporate multiple perspectives in decision-making and better consider the spill-over effects of policy actions and implications on each other's plans. Some of Singapore's notable involvements and efforts in preserving the cultural heritage-built attractions in Singapore include the following:

(1) SDG 8—Local engagement and public feedback were sought when revitalising the heritage precincts such as Chinatown Food through the efforts of Singapore Tourism Board and Chinatown Business Association. Similarly, in Little India, STB cooperated with LASALLE College of the Arts and the Little India Shopkeepers and Heritage Association to implement place-making initiatives such as the Deepavali Light Up and the ARTWALK Little India events. STB provides funding support via its Association Development Fund to tourism-related precinct associations for capability building and sustenance.

(2) SDG11—The Singapore government sets land aside for parks conservation despite land constraints. The 150-year-old Singapore Botanic Garden is Singapore's first UNESCO World Heritage Site. Ten per cent of land in Singapore is set aside for parks and nature conservation and today, more than 80% of households live within 400 m or a ten-minute walk to a park and expected to expand to more than 90% of households by 2030, by creating more neighbourhood and regional parks.[8]

[6] The primary task has been preserved in statements from UNESCO Recommendation on Historic Urban Landscape to UN's New Urban Agenda and OECD Conference on Culture and Local Development in 2016 and 2018, respectively. See United Cities and Local Governments (2010, November 17).

[7] UNESCO Institute of Statistics (2023).

[8] Ministry of Foreign Affairs Singapore (2018).

In spite of the complications of the built environment's contribution to the SDGs and the demands to put collaboration-based methods first, Singapore is determined to preserve its contemporary and heritage-built attractions. In particular, Singapore maintains the built attraction orientation while blending the design of past and new construction to bear Singapore's deep and unique cultural heritage. With increasing consciousness of the importance in sustaining Singapore's built cultural heritage, cultural heritage walking trails have been constructed, linking green spaces and under-and above-land heritage sites, to offer residents and tourists dedicated routes for walking and cycling without impacting nature and biodiversity.

In Table 4.2, a comprehensive summary is presented, highlighting the connections between the SDGs and Singapore's preservation initiatives for its built attractions.

Interlinking Cultural Heritage and Walkability in Singapore's Historical Ethnic Enclaves

Cultural heritage and walkability are crucial factors in shaping vibrant and liveable cities. Cultural heritage encompasses significant landmarks, historical structures, museums and other sites that contribute to a community's identity and sense of belonging. By integrating these cultural heritage sites into urban planning, cities can establish a unique sense of place and foster a strong community identity. This, in turn, enhances walkability as individuals are more inclined to explore and traverse areas that they find visually appealing and culturally engaging. Moreover, the presence of cultural heritage sites acts as a major attraction for tourists and visitors, leading to increased foot traffic, support for local businesses and job creation. This positive feedback loop creates a dynamic environment where the availability of cultural heritage sites continually attracts more visitors, propelling economic growth, vibrancy and further enhancing walkability.

The Singapore government acknowledges the cultural influence of the West on Singaporean society and emphasises the significance of establishing strong cultural identities by connecting individuals with meaningful places. In this regard, heritage landscapes offer a way to balance between Asian values and growing Western influences and preserve and strengthen the Asian heritage of its citizens by fostering a sense of rootedness and belonging. Singapore city is a genesis for being a natural

Table 4.2 Linking SDGs to Singapore's cultural heritage and built attractions

SDG		*Cultural heritage and built attractions*
1	No Poverty	• Vulnerable groups, such as seniors, have opportunities for employment within Singapore's heritage buildings, enabling them to attain the necessary funds for accessing essential services like water, sanitation and other necessities • The revitalisation of traditional crafts like Peranakan Nyonya kueh making, the preparation of diverse Indian and Chinese temple offerings and the development of walking tour guides can create employment opportunities and establish both traditional and contemporary professional portfolios • By promoting employment in fields like building reconstruction, local souvenir shops, museum guides and ethnic food services, cultural heritage enables communities to utilise their local skills and resources effectively. This empowers individuals with ownership and management control over land and heritage building premises, fostering economic growth and providing avenues for income generation. As a result, communities are able to uplift themselves from poverty and improve their overall well-being
2	Zero Hunger	• Through the reintegration and a common consideration of heritage fruition, it is feasible to curb the use of lands dedicated to preserve the national monuments, and at the same time to accommodate other food trades to feed the communities, including hi-tech agriculture and fish farming. The relation between new construction and land usage are explored by authorities to ensure a broader range of a mix of old trades and modern professions can continue to thrive and sustain livelihood and increase local food production • This initiative aligns with the goals of Sustainable Development Goal 2, which aims to promote sustainable food production systems, resilient agricultural practices and the preservation of ecosystems, while concurrently enhancing land and soil quality. While Singapore currently produces approximately 10% of its food domestically, the challenges posed by population growth and climate change necessitate a global effort to secure food supplies. To address this, Singapore has set an ambitious target to increase local food production to 30% by 2030 through the implementation of the '30-by-30" plan. This plan aims to foster self-sufficiency, boost productivity, and ensure the resilience of the food supply chain, thereby contributing to sustainable development and the protection of natural resources

(continued)

Table 4.2 (continued)

SDG		Cultural heritage and built attractions
3	Good Health and Well-Being	• Valuing cultural heritage serves as a vital source of nourishment for mental well-being, fostering stronger community bonds, enriching personal life journeys and nurturing a sense of identity as individuals connect with their heritage and its associated values • Museums such as the Asian Civilisations of Museum, National Museum of Singapore, former Saint Joseph's Institution (now Singapore Art Museum) and former Tao Nan School (now The Peranakan Museum) in Singapore undertake the role through education and preservation of cultural heritage • Connection with the heritage museums through walk, ride (on public transport) and cycle to these heritage monuments (e.g., museums) contribute to a healthy environment with less carbon footprint. By facilitating walking trails, individuals not only engage in physical exercise to maintain good health but also have the opportunity to spend quality time in heritage museums, fostering a deeper understanding of Singapore's multi-ethnic society and its rich heritage values
4	Quality Education	• National monuments and marked historical sites, such as Singapore museums, the former Nanyang University Library and Administration Building, College of Medicine Building and the former Raffles College (now NUS Campus at Bukit Timah), provide younger generations with immersive learning experiences beyond the confines of traditional classrooms. These sites serve as educational hubs, offering valuable opportunities for the Singaporean society to acquire enriching and high-quality education within the captivating walls of these esteemed built heritage monuments
5	Gender Equality	• Ancient buildings in Singapore, such as the Sri Mariamman Temple, Hajjah Fatimah Mosque, Thian Hock Keng (Palace of Heavenly Happiness) Hokkien temple, Jamae Mosque, Sultan Mosque and Former Siong Lim Temple (now Lian Shan Shuang Lin Monastery), showcase a distinct attention to gender-based segregation and the allocation of spaces. These architectural marvels serve as tangible reminders of the historical roles and societal positions of men and women

(continued)

Table 4.2 (continued)

SDG		Cultural heritage and built attractions
		• By examining these built cultural heritage sites, researchers gain valuable insights into the past lifestyles, cultural norms and traditions that have shaped gender dynamics. Exploring the contribution of heritage buildings to gender equality is an area of compelling research interest, as it sheds light on the progress made and the work that remains in promoting equality and inclusivity in modern society[9]
6	Clean Water and Sanitation	• The rebuilding of the National Aerated Water Factory, undertaken by the owner to commemorate the industrial legacy of Kallang River, involved the conservation and extension of the factory site in 2018. This transformation resulted in the coexistence of traditional water management systems, which hold historical and technological significance, alongside more modern techniques • Preserving the National Aerated Water Factory in its functional and unaltered state would serve as a powerful symbol of the resilience of heritage buildings, showcasing their ability to provide essential water resources. The sire has the potential to inspire innovative solutions for accessing clean water and implementing effective sanitation measures
7	Affordable and Clean Energy	• Integrating green and modern technologies into heritage buildings presents challenges in promoting the adoption of renewable energy sources. The introduction of the Net Zero Energy Building (NZEB) concept, mandated by the 2010/31/EU Directive on Energy Performance in Buildings, calls for further research on retrofitting existing historical structures, which constitute a sizeable portion of the built environment. These efforts aim to establish additional standards aligned with the NZEB concept • The B House, Singapore's first zero carbon house, is nestled within the spaces of colonial black and white bungalows from the British colonisation era. Pomeroy Studio, a Singapore-based design and architecture firm, accomplished this remarkable achievement in 2016 by incorporating key features of the colonial bungalows, such as large verandas, generous roof overhangs and adjustable shutters for natural airflow and sun protection. The installation of solar panels enables the generation of clean energy, making this home located in Bukit Timah an exemplary model for reduced water and energy consumption

(continued)

[9] See Dempsey et al. (2020), Hamoud (2007), Herbig et al. (2012) and Verster (2014).

Table 4.2 (continued)

SDG		*Cultural heritage and built attractions*
8	Decent Work and Economic Growth	• The economic growth and employment opportunities in a region are closely tied to the cultural and social capital present in the local community, making heritage buildings an attractive factor for businesses, creative industries, residents and visitors alike. Sectors such as building maintenance, urban regeneration, tourism, museums, archaeology and handicraft production directly benefit from the preservation of these buildings. Moreover, heritage-listed properties have the potential to enhance the overall real estate and land value of their surroundings • An excellent example of the economic significance of heritage buildings is Singapore's Botanical Gardens. Originally established as a British colonial botanic garden in 1859, it evolved into a centre for botanical and horticultural research, including the cultivation of rubber plantations. Today, the Gardens have transformed into a world-class destination for conservation, recreation and education, attracting both locals and tourists. The buffer zone surrounding the Gardens, known as "Good Class Bungalow Areas", maintains a low-rise and low-density environment, contributing to its appeal and generating business opportunities in the form of thriving cafes and eateries • Engaging the public and involving local communities in the restoration and preservation of precincts is crucial. In Singapore, collaborations between organisations such as the Singapore Tourism Board, Little India Shopkeepers and Heritage Association and LASALLE College of the Arts have led to successful place-making activities like the ARTWALK Little India event and Deepavali Light Up, revitalising the cultural district. Similar efforts have been made in Chinatown, where the Chinatown Business Association and Singapore Tourism Board have worked together to enliven the Chinatown Street Food scene. The Singapore Tourism Board provides funding support to tourism-related precinct associations, empowering them to become self-sustaining entities and fostering the growth and development of these industries

(continued)

Table 4.2 (continued)

SDG		*Cultural heritage and built attractions*
9	Industry, Innovation and Infrastructure	• Heritage buildings are spaces of ongoing innovation and experimentation from the past centuries, inspiring urban and building design professionals in modern times of their resilience and adaptability, while contributing to new creations or solutions to the present built environment that becomes the heritage of tomorrow • Four heritage hotel buildings in Singapore have undergone restoration, each with its unique charm and history. The Warehouse Hotel, originally a rundown godown from 1895 along the Singapore River, has been transformed into a modern, industrial-themed hotel with preserved tethers and beams, offering a blend of heritage and contemporary design in its 37 rooms and public spaces. The iconic Raffles Hotel, which dates to 1887, reopened in 2019 after an extensive two-year renovation, delicately refurbishing its interiors while maintaining its luxurious heritage. The Sentosa Barrack, a former British army barracks, was converted into a cosy hotel in 2019, preserving its vintage elements such as moulded columns and timber louvered windows and doors in all 40 rooms and suites. The Fullerton Building, originally the General Post Office built in 1928, was renovated in 2001 to become The Fullerton Hotel Singapore, featuring 400 stylishly contemporary rooms and suites. The restoration of this building paid careful attention to preserving its grand façade, porte cocheres and Doric columns, reminiscent of the Palladian architecture of the city-state
10	Reduced Inequalities	• The vibrant ethnic enclaves of Chinatown, Kampong Glam and Little India underwent a transformation in the 1980s, evolving from mono-racial settlements of Chinese, Malay and Indian communities into historic districts. Despite Singapore's reputation as a bustling urban landscape, these districts stand as living testaments to the rich cultural heritage that thrived prior to Independence, symbolising the nation's commitment to preserving and celebrating its diverse roots
11	Sustainable Cities and Communities	• Singapore takes pride in its modern architectural gems, including the renowned Singapore Sports Hub and the transformed CHIJMES, which were once the Kallang Stadium and Convent of the Holy Infant Jesus complex, respectively. These heritage buildings exemplify Singapore's commitment to sustainable urban development, integrating seamlessly with the surrounding environment and enriching the city's cultural and architectural landscape

(continued)

Table 4.2 (continued)

SDG		Cultural heritage and built attractions
		• A shining example of sustainable urban planning in Singapore is the Singapore Botanic Gardens, a UNESCO World Heritage Site. Situated just outside the bustling shopping district, this green oasis showcases the successful integration of nature within the urban fabric, promoting biodiversity, providing recreational spaces and contributing to the overall sustainability and liveability of the city
12	Responsible Consumption and Production	• Built heritage plays a crucial role in promoting responsible consumption and production by facilitating sustainable and innovative practices such as the adaptive reuse of natural resources, reduction of material, water, energy and soil waste and minimising demolition costs. These approaches align with the principles of responsible and efficient resource management while preserving the essence of vernacular architecture • Singapore's commitment to responsible consumption and production is evident in initiatives like the Active, Beautiful, Clean (ABC) Waters programme launched by the Public Utilities Board (PUB) in 2006. With 36 projects across Singapore, including the Alexandra Canal, the programme transforms canals, reservoirs and drains into scenic water bodies that serve as recreational spaces. The Alexandra Canal, with its historical significance and colonial era name, not only manages water but also provides a vibrant and picturesque setting for people to enjoy, showcasing the harmonious integration of sustainable water management and responsible urban development
13	Climate Action	• The recovery and preservation of heritage sites and practices contribute valuable knowledge and experiences in climate mitigation, even as local environmental conditions change. Sustainable models, such as walkable and mixed-use historical centres and energy-efficient construction of heritage buildings, equipped with energy-saving devices, play a crucial role in reducing carbon footprint. These efforts align with the goals of SDG7 (Affordable and Clean Energy) and SDG12 (Responsible Consumption and Production)

(continued)

Table 4.2 (continued)

SDG		*Cultural heritage and built attractions*
		• URA of Singapore has taken proactive steps to address climate change by transforming historic districts into car-lite shared spaces that prioritise pedestrians. Renowned quarters like Kampong Glam, Boat Quay, Chinatown and Little India have embraced these transformations. In these bustling social and commercial areas, side streets adorned with traditional shop-houses have been designated as car-free zones, particularly on Sundays since 2014. This periodic closure of selected roads, including Club and Ann Siang Streets in Chinatown, not only reduces carbon emissions but also creates an inviting atmosphere for al fresco dining and fosters social connections. These initiatives generate cherished memories and foster a deeper sense of pride in Singapore's rich cultural heritage
14	Life Below Water	• A notable example is the discovery of two shipwrecks near Pedra Branca, located 24 nautical miles east of Singapore. The first shipwreck yielded a fascinating collection of artefacts, including fourteenth-century Chinese ceramics, notably the exquisite blue-and-white porcelain from the Yuan Dynasty. Some of these artefacts can be traced back to the 1300 s. The second shipwreck, known as Shah Munchah, was an Indian-built merchant vessel that sank in 1796. The careful research and conservation efforts surrounding these shipwrecks have resulted in the curation and display of these artefacts in museums, allowing us to appreciate and learn from our underwater heritage while protecting the delicate marine environment
15	Life on Land	• An excellent example of this is the restoration of St James Power Station located at 3 Sentosa Gateway in 2018. The restoration project made use of local conservation expertise and involved extensive historical research and interdisciplinary skills. Careful attention was given to preserving the unique spatial quality of this industrial landmark, including the delicate implementation of floating floors and the creation of a central skylit atrium and communal space. The carpark area underwent a revitalisation with a redesigned landscape, a new lawn and a specially designed covered walkway. In a tribute to the building's history, an ingenious transformation of the chimney has turned it into a heritage gallery, showcasing the stories of former employees and the remarkable journey of this magnificent heritage building

(continued)

Table 4.2 (continued)

SDG		Cultural heritage and built attractions
16	Peace, Justice and Strong Institutions	• Heritage represents a shared resource that embodies community values, fostering partnerships between the public and private sectors to drive consensus in urban development processes. These initiatives also contribute to the strengthening of city administration and relevant institutions. The overarching goal is to leverage the symbolic value of heritage and create a sense of belonging among the public through urban regeneration projects • URA is actively involved in preserving historic districts like Chinatown, which form an integral part of Singapore's urban landscape. Beyond its nostalgic significance, Chinatown's intangible heritage plays a crucial role in shaping the collective identity of the nation. To ensure the preservation of this heritage, the government has collaborated with various stakeholders, including resident arts groups, grassroots leaders and business owners, to forge strong public–private partnerships. This collaborative approach has resulted in the introduction of three key policy proposals, namely the establishment of a statutory heritage trust, the promotion of the commercial viability of traditional trades and businesses and the revitalisation of Chinatown to appeal to the younger generation. These efforts aim to enhance the understanding, appreciation and sustainable management of Chinatown's heritage for the benefit of present and future generations
17	Partnerships for the Goals	

meeting point of sea routes due to its privileged location at the tip of the Malay Peninsula. Historically, Singapore prospered as a vessel trading post for Arab dhows, Buginese schooners, Chinese Junks and Portuguese battleships. Singapore's rich history has led to the preservation of various facets of its multicultural identity. These preserved elements encompass both tangible and intangible cultural heritage, offering a vibrant reflection of Singapore's diverse cultural tapestry. In 2021, EDB launched Singapore Unfolded! Culture and Heritage initiative with the help of two friendly guides, Jackie Goh and Natalie Kwee from 8EyedSpud, a creative studio. Visitors are introduced to the trifurcation of communities, religions, languages and Singapore's ethnic diversity through leitmotif scores of music and nostalgic past stories unique to each of the ethnic groups.

Little India has been identified as a significant area for conservation, recognising its cultural and historical significance in Singapore. The walking trails in the historic Little India provide an eclectic and buzzing space in Singapore. Little India was once home to the country's first race-course and houses Hindu temples, along with modern businesses and olden-day trades that offer groceries, jewellery, sarees and electronics to its visitors. The precinct has a few names due to its historic importance, *Kandang Kerbau* (or buffalo pen in Malay) and *Tekka* (means bamboo in Hokkien). The Indian Heritage Centre has five thematic galleries depicting the Indian origins, movements and migration to Singapore and it is acclaimed as the first South East Asian Museum that concentrated on the rich Indian community's heritage. Visitors can look out for the House of Tan Teng Niah in Rochor which was built in 1900. Tan ran a rubber smokehouse at Kerbau Road and a few confectionery factories in Serangoon Road. Jothi Store & Flower Shop is one of Little India's signature shops. The shop started from the humble beginnings as a five-foot way stall. Today, it fulfils the multitude needs of the Indian community especially during the Indian festivals.

Chinatown stands as a historic enclave that once served as a settlement for Singapore's Chinese immigrant population. It holds great cultural and historical significance, showcasing the rich heritage and traditions of the Chinese community. Each street in Chinatown has its own quaint identity of specific trades from hawkers, peddlers, businesspeople, traders, craftsmen, death-houses and delicacies. Rich artefacts of unskilled labourers (or coolies) are found which Chinatown was one of the transit points before moving to Malaya. The *kampong* lifestyle and the zone division were evident among the dialect groups. The Teochews occupied the Carpenter Street, Garden Street and South Canal Road. The Hokkiens were settled in Hokkien Street and Telok Ayer Street, while the Cantonese located at Temple Street. Today, one can walk and enjoy the culture and heritage monuments, artefacts and food in five districts of Chinatown (Kreta Ayer, Tanjong Pagar/Duxton, Ann Siang/Club Street, Telok Ayer and Bukit Pasoh). The walking trails house some iconic attractions, including the Thian Hock Keng Temple where early China immigrants prayed to the sea goddess Mazu in gratitude for their safe arrival to Singapore. The original construction of this temple did not use a single nail to build the monument and has a meaningful name that translates into the *Palace of Heavenly Happiness*. *Kong Chow Wui Koon* is a clan association and remains a custodian of Chinese heritage and preservation of the traditional arts. Visitors to the *wui koon* can learn about the dragon and lion head on its Kungfu floor.

Visitors to the *Kampong Glam* (*Kampong* refers to compound in Malay and *Glam* is a long-leaved paperbark tree) cultural heritage trails would get to learn about the past lifestyles of the Malay communities. The Glam provides materials for boat-making and producing food seasoning and medicine. Kampung Glam was known to be a strategic port town due to its location near the sea for commercial trading. Residents and tourists would find the area a blend of culture, lifestyle and history in their walking routes. The area remains a significant mark of its zenith when the Malay sultanate ruled and an important beacon for its Malay community. The historic district is home to Sultan Mosque, a majestic golden dome built by Sultan Hussein Shah in 1824. The mosque is located at 3 Muscat Street, and many are attracted to the mosque's grand golden domes decorated with glass bottle ends. The mosque is a symbol of solidarity for the Muslim community. Another monument is the two-storey *Istana* (Palace in Malay) Kampong Glam beside the Sultan Mosque. Built in 1843, the palace was home to Sultan Hussein Shah and his descendants. The building was designed by George D. Coleman, hence the palace bears classical European architecture elements such as pilasters of the Doric order. The monument also exemplifies traditional Malay architecture features such as constructing its second storey on stilts and its large-pitched roof characterising the Malay *Limas* or *pyramidal* House. The palace was a prime venue for tea parties and royal weddings hosted by the Malay community back then. The palace underwent refurbishment works from 1999 to 2004 and was converted into a Malay Heritage Centre by Prime Minister Lee Hsien Loong in June 2005. Today, the centre is one of the primary heritage buildings in Singapore to showcase the cultural traditions and rich arts of the Malay community.[10]

By navigating the STB website, visitors can dive deep into the district's history and make their walk to the Malay Heritage Centre that sits on the space of a former palace. Residents and tourists can glance the life of the seafarers and the Malay seafaring vessels when visitors participate in the cultural activities as well as walk in its six permanent galleries. Additionally, there are beautifully preserved royal artefacts in the galleries, reflecting the district's status as a regional printing hub. *Haji Lane* is the narrowest lane in Singapore but has much to offer to its visitors. It is a vibrant enclave with beauty salons, relaxing cafes and local fashion boutiques.

[10] National Heritage Board (2023b).

For example, one can find beautifully crafted jewellery at Silver Triangle, unique Singaporean trinkets at Craft Assembly, and elegant accessories and home décor at Hygge. Treasure hunters can make their walk to the Malay Art Gallery, a shopping gallery famous for its brooches, plates and *kris* (traditional daggers with wavy blades), Sifr Aromatics for self-concocted perfumes and Molkan Fabrics for textiles, silk and scarves. *Glam* Gallery is tucked away in a back alley where old meets new in Singapore's first outdoor gallery. While walking through the alley, visitors are in for a visual feast. There is an eclectic mix of contemporary arts works, close to 30 wall murals and works of street art by local and international acclaimed artists.

Singapore is actively engaged in the preservation of the Eurasian cultural heritage. There are around 15,000 to 30,000 Eurasians, accounting for less than 1% of Singapore's total population.[11] The tight-knit Eurasian community represents the descendants of sixteenth and seventeenth-century Iberian settlers who married local Malay women. A hybrid identity revolves around their own creole language (Papiá Kristang) and Roman Catholicism, while absorbing cultural influences of later arrivals from the Netherlands, India and China.[12] The Eurasian community called *Katong* as their home in Singapore from the early twentieth century. Restoration and conservation of heritage shop-houses, the preservation of iconic landmarks such as the Church of the Holy Family and Eurasian Community House, and the establishment of heritage trails and markers highlight the historical and cultural significance of the area. Walking tour trails within its galleries allow participants to take part in a Portuguese folk-dance workshop and learn its famous traditional Portuguese Malaccan folk song, *Jinkli Nona*. Others can learn how to make Eurasian delicacies such as traditional sugee cake and the super-spicy Devil's Curry in a hands-on cooking session.[13]

Other than Singapore's historic *ethnic enclaves* which draw the attention to its rich cultural heritage, over the years, more of its tangible and intangible cultural heritage settings are discovered, publicised and the cultural heritage sites, monuments and artefacts offer unique attractions for both tourists and residents. Cultural heritage walking trails

[11] Singapore Tourism Board (2023).

[12] Louie (2018, March 19).

[13] Singapore Tourism Board (2022).

in the *Garden City* has become a multi-function place for education, recreational and commercial purposes. Some of the most beautiful architectures in Singapore are the traditional vernacular buildings such as the plantation-style house in Burkill Hill, Tudor-style architecture at Chek Jawa and the Peranakan shop-houses in Koon Seng Road. The walking trails allow visitors to learn, immerse and visualise the nostalgic feelings, mood and glory. There are also historic bridges spanning the iconic river linking the Clarke Quay to Boat Quay, Civic District to Commercial District and more. Seven bridges were gazetted collectively as the nation's 73rd National Monument. Several facelifts were needed to keep the bridges in pristine condition.

Hotels, particularly those with historical or architectural significance, hold a special place in conservation efforts. These hotels often embody cultural heritage and serve as important landmarks. By preserving and repurposing old hotels, Singapore aims to protect the architectural integrity and historical value of these buildings, preventing their loss due to urban development or neglect. Raffles Hotel, an iconic landmark in Singapore, proudly holds the prestigious title of being one of the oldest hotels in the country. With its roots dating back to its establishment in 1887 by the Armenian hoteliers, the Sarkies Brothers, Raffles Hotel has stood the test of time, embodying the grandeur and elegance of a bygone era. Throughout its illustrious history, the hotel has played host to numerous notable figures, witnessed significant events and etched its name into the annals of Singapore's cultural heritage. Today, Raffles Hotel stands as a symbol of timeless luxury and charm, preserving its heritage while offering a remarkable experience to guests from around the world.

The Fullerton Hotel built in 1928 had withstood the WWII occupation and is Singapore's 71st national monument. The Capitol Kempinski Hotel Singapore built in 1904 and 1930 housed two historic landmarks; the Venetian Renaissance-style Stamford House (1904), built in the years leading up to World War I and the neoclassical-style Capitol Building (1930). The Capella Singapore was built in 1880s and used to accommodate the British officers of the Royal Artillery and their families. Goodwood Park Hotel was originally built in 1900 as an elite enclave called the Teutonia Club for the German community in Singapore but was converted into a hotel by 1929 which received some of the world's esteemed guests such as the Prince of Wales of England and Duke of Windsor at the end of the 1930s. The hotel later became the residence for high-ranking Japanese soldiers, a total of three years from 1941 during

WWII and then as a British War Crimes Court before returning to the new owner of a Manasseh descent, Mr. Vivian Bath in 1947. The Barracks Hotel in Sentosa built in 1904 boasts a long history. The two-storey buildings were once the *Blakang Mati* Artillery Barracks, a prisoner-of-war (POW) camp during WWII. Hotel Ford Canning was opened in 1926 and once served as the Administration Building of the British Far East Command Headquarters until 1942 before being taken over by the Japanese Military during WWII. The building was also occupied by the Singapore Armed Forces when the nation gained independence in 1965 and was utilised by the Singapore Command & Staff College.

Visitors on foot in City Hall would not miss the striking Gothic exterior of the St Andrew's Cathedral. The edifice was consecrated in 1862 and was built with the hands of Indian convict labourers. The cathedral isle resembles the small-town English church halls. In the vicinity is *Chijmes*. Formerly a Convent of the Holy Infant Jesus and a campus for a renowned Catholic girl's school, *Chijmes* is now a popular place to host wedding events, an enclave that has much to offer with its relaxing ambience found in its restaurants, bars and cafes. The other monuments in the City Hall area worth exploring on foot are the Victoria Concert Hall and Victoria Theatre which have hosted world-class performances and performances by Singapore's Symphony Orchestra. The concert hall is the symbol of the nation's growing performing arts scene. However, both have their share of historical stories. It was a makeshift hospital during the Japanese air raids in 1942 to treat casualties and the Victoria Memorial Hall was where trials of war criminals were conducted. The Sun Yat Sen Memorial Hall at Tai Gin Road was once the nerve centre of Sun Yat Sen's revolutionary movement in Southeast Asia. Teo Eng Hock who was the owner, and a rubber magnate offered his two-storey villa to support Sun's revolutionary activities to defeat the Japanese invasion of that era. The memorial hall is preserved to commemorate Sun's legacies. Its galleries hold artefacts contributed by the Chinese community, with 3D mapping and hologram projection features.

The Battle Box is another spot that walkers can head to. This space was important to the Malaya Command, which was built between 1936 and 1941, as their head command operations. Visitors can walk through the hour-long tours of the underground command centre with replicas in the genuine rooms where military officers were stationed there. The docents re-tell the events that took place during the defeat of Malaya and Singapore in the hands of the Japanese. Not far away, sits St James Power

Station at the Harbourfront. Visitors can take a stroll and relax in the venue, famous for nocturnal visitors that are looking for nightlife entertainment. It was one of Singapore's oldest power stations that supplied electricity to shipyards and surrounding residential neighbourhoods.

Lovers of historical artefacts, cultural relics, rare masterpieces and thought-provoking paintings would find the Singapore's *museum walking trails* interesting. Singapore is a flourishing cultural centre and has hosted big-name museum exhibitions over the years. The largest visual art gallery in Singapore showcasing mostly local and Southeast Asian pieces is the National Gallery, which is the largest national museum consisting of two monuments, the City Hall and the former Supreme Court, occupying 64,000 square metres of floor space. The gallery hosts more than 9,000 items and is open to the public. Its collection focuses on the relationship of Singapore's culture with other Asian cultures and the world through a variety of media. The Asian Civilisations Museum is home to more than two thousand artefacts from the past civilisations of West Asia, Southeast Asia, China and South Asia. A walking distance from the City Hall station, lies the Central Defence Heritage Gallery which now is a museum with fire trucks and equipment open to the public. Its red-and-white building construction was completed in 1908 with a living quarter for firefighters and a watchtower. The monument used to monitor the civil defence developments of the country. The nation's oldest museum, the National Museum of Singapore, opened its doors in 1887. The permanent galleries narrate the city developments over the years, and a must-visit venue. Those who are taking a walk at the Bayfront Avenue can hop into Art Science Museum which houses remarkable works and exhibits from prominent artists including Leonarda da Vinci, Salvatore Dali and Andy Warhol.

Moment of Imagination and Nostalgia with Toys (Mint) at Seah Street houses a toy collection worth over five million, and up to 50,000 vintage playthings from different eras, including the pre-WWII epoch. It is the world's first built museum for toys. The Trick Eye Museum located at Waterfront, Resorts World Sentosa offers a sci-fi treat whereas the Red Dot Design Museum at Marina Boulevard displays mind-blowing product prototypes, innovation concepts and communication design. The Singapore Musical Museum Box at Telok Ayer Street showcases forty antique musical boxes from the nineteenth and twentieth centuries. The museum was established by Naoto Orui, a renowned Japanese collector. One of the highlights is the rare Edison Opera phonograph which comes with a wooden horn handcrafted by Thomas Edison himself.

Reflections of Bukit Chandu at 31-K Pepys Road hosts unique artefacts and multimedia works that would take one on a trip to the past battles of brave soldiers who fought to preserve their Motherland. The Vintage Cameras Museum Singapore at Jalan Kledek is the world's biggest camera-shaped building set up by two cousins, AP Shreethar and Solaiyappan Ramanathan. The camera-shaped building houses vintage photography equipment and an enormous collection of over a thousand vintage cameras. Singapore Drive is home to the Lee Kong Chian National History Museum at the National University of Singapore. Wildlife lovers can view pre-historic dinosaur fossils, plants and fungi, sea creatures, skeletons of giant mammals and the evolution of plants and habitats in the museum's specially designed zones and gardens. The Changi Museum at Upper Changi Road North houses the artefacts of prisoners-of-war and serves as a memorial to the victims and soldiers of WWII. Peranakan Museum at Armenian Street is a picturesque three-story shop-house and displays a wide range of Peranakan art, jewellery and furniture that reflects the rich lifestyle of this community. Private art galleries play a vital role in nurturing and showcasing individual art collections, providing a platform for artists to exhibit their works and connect with art enthusiasts. A remarkable example of this is the Gillman Barracks located at Lock Road, which has transformed a colonial barracks into a vibrant art enclave. Here, contemporary art takes centre stage, with a diverse range of art galleries displaying innovative creations. Another notable artistic hub is the Goodman Arts Centre, nestled within the culturally significant Mountbatten zone. This dynamic centre serves as a creative hub, offering shared facilities for artists and fostering collaboration among various artistic groups. These artistic spaces not only contribute to the local art scene but also enhance Singapore's cultural landscape by providing accessible avenues for the appreciation and celebration of art.

Singapore's Challenges in Maintaining Its Built Attractions

In the early 1960s, Singapore was aware of the need to preserve the city's historic sites in spite of the random building/structure demolition. Two United Nations member teams, one led by Erik Lorange and another by Charles Abram offered advice to the Singapore government with regard to its heritage preservation efforts. In particular, the notion of demolishing old buildings for the sake of urban renewal was abandoned and replaced by *conservation, rehabilitation and rebuilding* values. The PAP with the Lorange-Abram teams from United Nations readily accepted the need to kickstart a new project under a Master Plan to break colonial continuity and remake Central Area. Housing was a major focus because many urban kampungs and existing shop-houses were in a bad state. The UN teams recommended a massive building programme to cater to 10,000 public housing every year for the next twenty years. Between the two teams, Lorange was commissioned by the UN Technical Assistance Administration. In Lorange's report, he supported any means not to wholesale remove the old buildings in the Central Area. Lorange encouraged comprehensive redevelopment along *modern lines* to prepare for Singapore's growth while rehabilitation acts as a temporary resort. Urban renewal would take many years and Lorange suggested encouraging private owners to undertake *limited modernisations, repair and rebuilding* for a small number of properties while in occupation. Lorange's action plans reinforced the HDB's perspectives that urban renewal was fundamentally "slum clearance" and "renewal".[14]

There are always new buildings constructed and making their way into Singapore's urban landscape. Thus, efforts are put in place to preserve these iconic heritage markers for Singapore's future generations to appreciate for a long time to come. However, balancing the economic versus non-economic value of preserving the cultural heritage building and sites, remains a challenge. More so when it comes to assessing the asset value of cultural heritage building and sites, while realising the tangible results oriented to the United Nations Sustainable Development Goals (SDGs). While cultural heritage has been identified as an enabler of sustainability and an authentic driver for sustainable development, in the case of Singapore, the role that cultural heritage plays in meeting SDGs in the Agenda 2030 has more to be explored. It is encouraging to note that Singapore has made significant advancements in incorporating cultural heritage sites into the city's infrastructure.

[14] See Dobbs and Loh (2023).

Concrete Actions Supporting SGDs

Singapore had rolled out concrete actions to preserve the national cultural heritage building and sites that can best support the fulfilment of SGDs. Profoundly interwoven, tangible and intangible heritage conceptualisations are tipped as *message* and *medium*,[15] representing the idea of buildings as *socio-material hybrids*.[16] Tangible and intangible heritage can weave exceptional lived experiences for Singapore's community of its past, present and future.[17] Modelling the conservation and management of the built heritage as cited by Rachel Douglas-Jones, John Hughes, Sian Jones and Thomas Yarrow "*...is a complex process involving not only physical fabric, but also cultural, aesthetic, spiritual, social and economic values*".[18]

Singapore has been channelling efforts to build an inclusive society that in tandem fulfils the SGDs. To enable good access including people with disabilities and its ageing population, Singapore's URA and LTA had been the main government bodies in planning and realising the following plans. Tucked within a heavily urbanised city of a small land area of 709 square kilometres, Singapore's contemporary and heritage buildings, are located within walking distance from bus stops and/or train stations.[19]

Other measures have been undertaken while supporting preservation of Singapore's cultural heritage buildings and sites. Adaptive reuse in restoration and repurposing, and the recovery of underused buildings as new tourist attractors are some examples. This practice is associated with Singapore's adoption of the circular economy in the tourism sector. The circular economy mode directly avoids the resource usage needed for construction of new buildings, forming the tertiary building sector. Recovery of cultural buildings and cultural sites had lessened the need for construction of new construction for cultural halls, offices, schools and service centres. Adaptive reuse work driven by Singapore's circular economy requires communities' participation, legal status settlement and asset identification that must be solved before reuse actions are undertaken.

[15] Taylor (2015).

[16] Djabarouti (2020).

[17] Kearney (2008).

[18] Douglas-Jones et al. (2016).

[19] Trading Economics (2022).

Coupled with the nationwide movement to enable Singapore's inhabitants and tourists to walk, cycle and ride, there are obtainable benefits, including uplifting of the local economy with new businesses, new jobs, revenues from local expenditures and tax, revitalisation of Singapore's rich culture and creation of attractive and vibrant environments and communities for people to thrive in.[20] Furthermore, centred on the concerns on SDG11—Sustainable Cities and Communities, it seems fitting to remind ourselves of the definition of reuse purposes more so in enabling walkability for tourists and instil in the daily lives of locals, and in the entirety of the city's urban functions. Reuse of this kind can reconcile locals with the tourist fulfilment of urban spaces. Enabling walkability in the recovered use of heritage buildings indoor and surroundings allow both locals and tourists to enjoy the culture, arts and heritage of Singapore. Promoting walkability can suppress the needs to form new spaces for new urban functions and the accompanying negative environmental impacts. More importantly, despite the attention to tourists, it is also to make known to locals of the concrete visibility in joint rather than exclusive use of heritage assets—*heritage is meant for everybody*. With a slower pace of walking through the heritage buildings and sites, one can get to rediscover the sites' original meaning—neglected or lost in time and reinforce the locals' sense of belongingness to their ancestry, which can trigger more appreciation and active involvement to preservation and maintenance initiatives.

Singapore's heritage buildings had been able to inspire visitors that spells the nation's rich traditional lifestyles and ancient practices. More tender care towards resource usage, mingling together with Singapore's deep cultural heritage values, is posed as a more delicate approach. From the singular environmental viewpoint, promoting walkability would lessen the conflict of the use of green spaces in maintaining Singapore's community psychological welfare and microclimatic comfort which can be at odds with soil utilisation and high-density principles. Singapore launched its Active, Beautiful, Clean (ABC) Waters programme in 2006. Lakes, streams, rivers, reservoirs, canals and drains add on as recreational spaces that encourage walkability and improve locals' well-being. Other than preserving some of Singapore's oldest canals, the most iconic is the

[20] Gravagnuolo et al. (2017).

Alexandra Canal, a heritage monument to channel water for Singapore's use from Johor Bahru as far as back in 1945.[21]

As evident, long-term physical planning had made the Singapore city more liveable including designing more pedestrian-friendly spaces. One can bask in the quaint attractions of its historic ethnic enclaves, cultural heritage walking trails in its garden city and get entertained. Endless possibilities await Singapore residents and tourists by curating their own walking trails that acquaint them with a kaleidoscope of Singapore's culture and heritage. Historical enclaves in Joo Chiat/Katong, Little India, Chinatown, Kampong *Glam* have so much to offer from the kampong lifestyle, cultural arts, ethnic food and the religious shrines, mosques and temples that devotees can go to pray. A walker is taken back to the past and gets to unravel stories that each monument brings from bridges to majestic hotels, houses, museums, hospitals and correction facilities. Withstanding the test of time, some of these monuments have their own dark history to tell, while others express cultural heritage pride, art celebration, uniqueness in architecture and modern sophistication.

The Green Plan 2030 expresses Singapore's commitment to take concrete actions to construct a sustainable future and address climate change. The Green Plan is hailed as a *living plan* which will evolve with the Singapore society. As sustainability efforts gain traction in the global economy, there is a pressing need for Singapore to achieve its Green Plan 2030 targets towards a low-carbon future while conserving its contemporary and heritage-built attractions. Singapore government actions undertaken from the perspective of the SDGs have been applying adaptive reuse of heritage buildings and sites. These new spaces often focus on possible environmental implications, on how to accommodate for tourist purposes and on single-asset logics. Another key conduit for actions to stimulate walkability and preserving Singapore's built heritage can be branded in tourism and usage of its built environment. Also known as a *Trojan Horse*, the built heritage identified in Singapore's tourism landscape, can act alongside the socioeconomic and environmental lines for fulfilling wider benefits to locals and tourists.

[21] Kadisky (2017, April 6).

References

Dempsey, K., Gilchrist, R., Ashbee, J., Sagrott, S., & Stones, S. (2020). Beyond the martial façade: Gender, heritage and medieval castles. *International Journal of Heritage Studies, 26*(4), 352–369.

Djabarouti, J. (2020). Listed buildings as socio-material hybrids: Assessing tangible and intangible heritage using social network analysis. *Journal of Heritage Management, 5*(2), 169–190.

Dobbs, S., & Loh, K. S. (2023). The origins of urban renewal in Singapore: A transnational history. *Journal of Urban History, 49*(1), 60–84.

Domrongchai, A. (2022, September 28). *NYC gets its first Hawker Center—With vendors straight from Singapore.* Food & Wine. https://finance.yahoo.com/news/nyc-gets-first-hawker-center-164009557.html. Accessed on 1 January 2023.

Douglas-Jones, R., Hughes, J. J., Jones, S., & Yarrow, T. (2016). Science, value and material decay in the conservation of historic environments. *Journal of Cultural Heritage, 21*, 823–833.

Government of Singapore. (2023a). *Preservation of Monuments Act (CHAPTER 239).* https://sso.agc.gov.sg/Act-Rev/PMA1970/Published/19870330?DocDate=19870330. Accessed on 8 January 2023.

Government of Singapore (2023b, January 11). *Certificate of Entitlement (COE).* https://onemotoring.lta.gov.sg/content/onemotoring/home/buying/upfront-vehicle-costs/certificate-of-entitlement--coe-.html#:~:text=Understanding%20COE,-All%20vehicles%20in&text=To%20register%20a%20vehicle%2C%20you%20must%20first%20place%20a%20bid,exercises%20conducted%20twice%20a%20month. Accessed on 26 February 2023.

Gravagnuolo, A., Girard, L. F., Ost, C., & Saleh, R. (2017). Evaluation criteria for a circular adaptive reuse of cultural heritage. *BDC. Bollettino Del Centro Calza Bini, 17*(2), 185–216.

Hamoud, M. (2007). The old town of Zanzibar and the struggle for reclaiming women's space in world heritage. *World Heritage and Identity: Three Worlds Meet*, 27–38.

Herbig, U., Weichart, G., & Breuling, J. (2012). Gender roles and women's influence on traditional architecture and its development in Indonesia: a case study from Nias. In *Research & heritage: Research papers on architectural heritage* (pp. 263–287). The Saudi Commission for Tourism and Antiquities.

Kadisly S. (2017, April 6). Hidden waters blog. *Alexandra Canal Singapore.* https://hiddenwatersblog.wordpress.com/2017/04/06/alexcanal/#:~:text=Alexandra%20Canal%20makes%20its%20appearance,consort%20of%20King%20Edward%20VII. Accessed on 6 December 2022.

Kearney, A. (2008). Intangible cultural heritage: Global awareness and local interest. *Intangible heritage* (pp. 223–240). Routledge.

Louie, J. (2018, March 19). *Understanding the culture of Malacca's Kristang community.* https://www.destinasian.com/blog/news-briefs/malacca-kristang-community. Accessed on 1 January 2023.

Ministry of Foreign Affairs Singapore. (2018). *Towards a sustainable and resilient Singapore.* https://sustainabledevelopment.un.org/content/documents/19439Singapores_Voluntary_National_Review_Report_v2.pdf. Accessed on 8 January 2023.

National Heritage Board. (2023b). *Istana Kampong Gelam.* https://www.roots.gov.sg/places/places-landing/Places/national-monuments/istana-kampong-glam. Accessed on 1 January 2023.

National Library Board, Singapore. (2019, July 31). *To wreck or to recreate: Giving new life to Singapore's built heritage.* https://biblioasia.nlb.gov.sg/vol-15/issue-2/jul-sep-2019/preservation-buildings. Accessed on 29 December 2022.

SG101. (2022). *Preserving a Unique national identity.* https://www.sg101.gov.sg/infrastructure/urban-planning/unique#:~:text=In%201971%2C%20the%20Government%20passed,eight%20buildings%20under%20its%20protection. Accessed on 29 December 2022.

Singapore Tourism Board. (2022). *Eurasian Heritage Centre.* https://www.visitsingapore.com/en_my/see-do-singapore/culture-heritage/heritage-discovery/eurasian-heritage-centre/. Accessed on 1 January 2023.

Singapore Tourism Board. (2023). *Faces of Singapore.* https://www.visitsingapore.com/travel-guide-tips/about-singapore/people-of-singapore/. Accessedon 12 January 2023.

Taylor, J. (2015). Embodiment unbound: Moving beyond divisions in the understanding and practice of heritage conservation. *Studies in Conservation, 60*(1), 65–77.

Trading Economics. (2022). *Singapore—Land Area (sq. KM).* Available at: https://tradingeconomics.com/singapore/land-area-sq-km-wb-data.html (Accessed on 14 January 2023).

UNESCO Intangible Cultural Heritage. (1992–2022). *Hawker culture in Singapore, community dining and culinary practices in a multicultural urban context.* https://ich.unesco.org/en/RL/hawker-culture-in-singapore-community-dining-and-culinary-practices-in-a-multicultural-urban-context-01568. Accessed on 1 January 2023.

United Cities and Local Governments. (2010, November 17). *Culture: Fourth pillar of sustainable development*, (Statement). https://www.agenda21culture.net/sites/default/files/files/documents/en/zz_culture4pillarsd_eng.pdf. Accessed on 19 November 2022).

Verster, M. (2014). *The wall and the veil: reclaiming women's space in a world heritage site* (Doctoral dissertation, University of Pretoria).

CHAPTER 5

Gastronomy and Intangible Cultural Heritage

Gastronomy and food are commonly accepted as a primary showcase of tourism destinations. Not only for their appeal to patrons but due to their potential in increasing local business activity and enhancing the image of the destination. In many destinations all over the world, food heritage tourism, culinary experiences and personalised food tourism itineraries have emerged in offering a memorable and meaningful sense to appreciate the local food scene.

Gastronomy and walkability are intertwined in creating a thriving culinary scene within a walkable city. Walkable neighbourhoods with pedestrian-friendly streets and pathways make it convenient for food enthusiasts to explore a variety of gastronomic offerings. The proximity of restaurants, cafes and food markets in walkable areas enhances the accessibility and diversity of culinary experiences. People can easily stroll from one food establishment to another, discovering local flavours and engaging with the community. The combination of gastronomy and walkability fosters a vibrant food culture, attracting both residents and tourists to indulge in the culinary delights of the city.

This chapter explores food traditions, food culture and gastronomy-walkability connectivity in Singapore. Food tourism which has been imbued into the Singaporean culture offers an exceptional gastronomic sensation but also signifies the destination's cultural identity.

Melissa Liow Li Sa and Sam Choon-Yin, *Sustainable Urban Development in Singapore*, Sustainable Development Goals Series,
https://doi.org/10.1007/978-981-99-5451-3_5

Governmental and Non-governmental Gastrodiplomacy

Gastrodiplomacy was mentioned in an article, *The Economist* back in 2002 on how food becomes a self-promotion tool for Thai companies. This term has now evolved to something more intricate, multidimensional and indispensable.[1] *Governmental gastrodiplomacy* has been referred to as forms of concerted public relations efforts and investments by states and federal governments, often in partnership with non-state entities, to leverage a nation's standing and value through food. Food is becoming compelling means for accomplishing diplomatic goals. For superpowers and legitimising sub-national powers as well as to gain hard power for middle power nations with less global hard power assets. Gastrodiplomacy traverses across the diplomatic continuum rather than just being a niche of diplomacy.[2] The purpose is to establish emotional connections, and these gastrodiplomacy campaigns help in expanding the cultural influence overseas for many nations. The traditional role of gastrodiplomacy of foodservices has been maintained till the present days. Marketing communications are initiated for economic reasons to promote food products internationally, though such promotion also helps in spreading cultural influence. It is deemed the softer side of public diplomacy for many nations.

Non-governmental gastrodiplomacy underpins the view that food is a strong mechanism for bringing people together. Having a plain meal in the company of one another can spark voluntary communication and enjoyment among peers. Beyond hospitality, people face less prejudices and more friendliness and willingness to participate in shared experiences.

Gastrodiplomacy Versus *Gastronationalism*

When food is an aspect of division, the concept of *gastronationalism* often appears. This concept was coined by Michele Antonio Fino, a professor at *Pollenzo* whose research interests are in ancient rights and Roman Law. According to Fino, food creates and sustains the emotive power of national attachment and must be promoted among the same people,

[1] *The Economist* (2022, February 21).

[2] Rockower (2012) and Clarke (2016).

rather than only between people, citing how food has helped to reconcile factions and strengthen social capital after civil wars.[3] Food being a fundamental ingredient of human subsistence has become a conduit for interstate relations and diplomatic practice. It is a channel for communicating culture, discharging influence and sending out messages that portray goodwill or hostility.[4]

Food has become a subset of the soft power of a country that helps develop relations internationally while portraying a positive image of the country. Culinary diplomacy is mostly carried out with food as the core diplomacy aspect during banquets. Customs and traditions of a country are revealed much more in these food settings, including the Quirinal Palace menus. French gastronomy was regarded as the mandatory event for state lunch and dinner settings. Michele Antonio Fino, Anna Claudia Cecconi and Andrea Bezzecchi opined *gastronationalism* as no more than an admissible face of a suprematism. Chauvinism and discrimination even with regard to recipes and dishes are no longer tolerated because it is incongruent with the present-time idea of civilisation. Over time, the French cuisine has included Italian food, and ever since pasta has become a constant presence in these food settings. The food scene has changed to one that of colonial seal which overshadows the other food cultures, but at present, potentially solidifies local gastronomic identities. Cooking becomes a site for encounters and a place of exchange among populaces, exhibiting love for a single dish. Kitchens now serve as an Intangible Cultural Heritage of Humanity. Despite the increasing moderation of food choices, there is a gradual digestion of lionising locavorism which dignifies food origins and its direct local food provision. Ecological manifestations imply to yield conservative communitarianism and culinary nationalism. There is an emerging deglobalisation of eating manners. When people travel, they get to expose and try signature local food from various parts of the world.

[3] Cannarella (2022, June 8).

[4] Chapple-Sokol (2013).

Gastrodiplomacy in Tourism

Traditionally food has been appealing to societies across geographical spaces and cultural boundaries. Aeons ago, the ancient trade routes were centred on exchanging food products such as coffee, spices and sugar. Tourism is another occurrence that connects nations and people, subsuming the role of developing national identity. Gastronomy in tourism involves state and non-state entities that seek to pursue a positive national brand associations among foreign publics, through which tourists get exposed to national cuisine. Tourists are motivated to travel to the country as attracted by the potential positive experiences that they are about to experience of the national culinary brand.

In gastrodiplomacy in tourism, *food, diplomacy*, and *tourism* are interwoven in forging and maintaining a national brand, such that each of these components can only be understood and appreciated when establishing a picture of its relationship with the other two. As an illustration, *diplomacy* initiated via gastrodiplomacy campaigns compel institution and cuisine or *food* quality control of ethnic foodservices overseas, building positive links or *diplomacy* among patrons. Motivated to travel to the country or *tourism,* tourists visit *food* or culinary sites and restaurants, while gathering in-depth awareness and exposure to the culture and enhance *diplomacy*. Stimulating repeat visits to the nation falls in the tourism realm which in turn triggers repeat visits to ethnic restaurants overseas.

On performing cultural diplomacy through the premediated use of food cuisine to induce a nation's perceptions, it is apt to draw upon a middle-power nation such as Singapore. Singapore's local and global campaigns portray its famous hawker culture which carries the UNESCO's *List of the Intangible Cultural Heritage of Humanity* title. Singapore is the first Southeast Asian country to pursue formal and international food-based recognition. It is obvious with the nation's programmes and rather hardy branding efforts in gastrodiplomacy. On top of anything else, this middle power nation quest for the UNESCO's bid is propelled by a yearn, in cooperation with non-state actors, to boost the value and the positioning of its national brand through its food culture. Original as it can be, Singapore's bidding was not only underpinned upon national identity creation scoped within the country's borders but also a practice in positioning the country's unique propositions that segregate from its other nation states. This enables Singapore to secure the observable benefits of a tangible and emotional transnational relationship with foreign public officials. Simply, through its local food in its historic ethnic enclaves, and a distinctive feature on its gastronomic *kopitiam* and hawker culture. While encompassing local food diners' gastronomic experiences, a composition of food, culture and country image in Singapore is formed in alignment with the UNESCO Creative Cities of Gastronomy aspects that include (1) culture food outlets and restaurants, (2) education and training, (3) food activities and (4) local ingredients and cooking skills.[5]

Residents, tourists and foreign public officials at Little India Arcade will learn about early Indian settlers' enterprising spirit, enduring spirit of resourcefulness and commerce that thrive amid the Indian business community until today. Banana Leaf Apolo, another popular destination, is an eatery famous for its wide range of authentic Indian food, from vegetarian, curries and seafood dishes that are served on banana leaves.

Keong Saik Road and Bukit Pasoh located in Chinatown were formerly red-light district areas and had undergone a series of transformations. The places are today home to art spaces, hawker gems, trendy cocktail bars and Michelin-starred restaurants. One of the food highlights is to stop for a bite (or *yum cha*) that offers a dining experience in the modern-day tea house, with *dim sum* in food baskets on push carts and retro marble tables at 20 Trengganu Street.

Kampong Gelam is once home to Malay royalty and the country's oldest urban quarter. When strolling around the neighbourhood, foodies can find both local and global dishes to serve one's cravings for new flavours. For a traditional fare, stop for a bite in one of its well-known authentic family-run eateries, *Hjh Maimunah* Restaurant at 11 Jalan Pisang. The Michelin Bib gourmand listed eatery offers over 20 types of kueh and 40 authentic Malay dishes such as *ayam bakar* (grilled chicken in Malay) and *tahu telur* (beancurd omelette with spicy peanut sauce) in Malay. Some *must-try* dishes by foodies in this eatery include *murtabak* (pan-fried bread stuffed with eggs), *nasi padang* (steamed rice with several dishes) and their very own Zam Zam Singapore—*roti prata*, a flatbread that originates from South India. For foodies who hanker for global flavours, one can walk along the Middle Eastern fare at Bussorah Street, eat Swedish food at Fika Swedish bistro and café as well as Italian

[5] Cheng (2017).

food at Tipo Pasta Bar. Other than a food paradise, the enclave boosts a trove of retail gems and where trendy modern retail and traditional tradesmen eke their living. At nightfall, the place comes alive with bars with unique themes and live music spots. Music and merriment after dark, and one can quench their thirst at popular dinner spots for hungry visitors. Recommendations include trying the local-inspired cocktails at Piedra Negra, Ah Sam Cold Drink Stall and the Singapura Club.

Brand Resonance Model

Kevin Lane Keller's brand resonance model offers a useful theoretical framework to describe Singapore's gastronomic experience. A layer is developed upon the preceding layer to respond to a specific question about the branding that people can connect to. The first layer fulfils the question—*Who are you?* Or connects to the brand's identity (salience). The second layer fulfils the question—*What are you?* Or to comprehend the brand's meaning (performance and imagery). The third layer allows us to understand—*What about you?* i.e., *response*, when people know who and what the brand is about (judgements and feelings), and the fourth (the highest) layer establishes the relationship between the brand and people or *What about you and me?*, otherwise brand *resonance*.[6] The model would be suitable for exploring Singapore's gastronomic experiences in its historic ethnic enclaves, *kopitiam* and hawker culture. These facets of the country, in turn, encouraged the love for its food outlets, restaurants, *kopitiams*, local ingredients and cooking skills[7] while basking in its deep historical trails, loyalty forms for these notable cultural icons.

The model interrogates how country image impacts progressive brand dimensions to sustain a successful and equitable brand.[8] It makes sense in applying the brand resonance model and the concept of country image in a gastronomic setting to appreciate how local diners perceive the image of Singapore regarding its culture and gastronomy, and how far they would go to promote their national culture at the international level.

[6] Keller (2001).

[7] Kim et al. (2022).

[8] On the application of Keller's model, see Azadi et al. (2015), Jung et al. (2014), and Pappu et al. (2007).

Local culinary culture is a formidable means that assume the task of gastrodiplomacy, representing place branding.[9]

In a study undertaken by Bona Kim and her associates, four dimensions of a Creative City of Gastronomy were significant factors in the Singapore gastronomy tourism setting; (1) culture food outlets and restaurants, (2) education and training, (3) food activities and (4) local ingredients and cooking skills, exclude traditional food industry and markets, and regard for the environment and marketing of sustainable local products. This infers Singapore lacks awareness about the sustainability of local products and environment as well as authentic food markets for the traditional food industry. It is gathered that their awareness of Hawker Culture was not related to loyalty (towards Hawker Culture) in Bona Kim and associates' study. Henceforth, verifies that the brand resonance model is organised on a branding ladder in stages and of a particular order. The first stage is brand salience, which needs to be established before forging the second stage of brand love and this has to take place before brand loyalty is firmly established. The *void* is braced by the aggressive launch of Singapore's Green Plan 2030. Concrete targets for sustainability have expanded to culture and food (gastronomy). The Singapore government plans to develop a robust local community for both supply and demand sides. For instance, sustainable development movement must adequately link to sustainable choices of daily food consumption of locals and locals to be more involved in responsible production. Therefore, food entrepreneurs in Singapore are advancing local food sourcing and precision urban farming for local food markets.

Singapore takes immense pride in being the pioneering Southeast Asian nation to boast Michelin starred restaurants and food stalls. This includes the hawker food in Mod-Sin (Modern Singapore) themed outlets which has from one-to-three Michelin stars, the French world-renowned restaurant ratings. Singapore also offers all kinds of food activities for its city locals from cooking events, gourmet festivals, as well as food and cuisine-related events. From the mid-1990s, World Gourmet Summit and Singapore Food Festival (SFF) are annual events organised by STB. Multicultural local cuisines are publicised in these key events to reach out to both tourists and residents. Live master classes and online virtual tours themed *Rediscover the Foodie in You* was promoted in SFF 2020

[9] Gordin and Trabskaya (2013).

due to the pandemic COVID-19. The pandemic did not stop food lovers and festival goers to pursue forms of gastronomic delights in their homes despite the social distancing. These events can upraise Singapore's standing among locals and position the country as a prima gastronomic city.

Based on the underpinning brand resonance model in Kim's study, Singapore's food restaurants and outlets is most significant in influencing all three dimensions of the brand resonance model—the awareness of Hawker Culture, love of Hawker Culture and loyalty to Hawker Culture. The awareness is linked to the local ingredients and local cooking skills factor. The awareness and love are associated with the education (cognitive aspect) and training (affective aspect) factors. It is apparent to Singaporeans; the hawker centre is the most common food and beverage outlet that they patronise. This trails back to the deeply entrenched culture rooted in Singapore's history. From the mid-1800s with the arrival of migrant workers, street food culture has become a primary element of Singaporean gastronomy culture that represents multicultural aspects. Locals regard Singapore hawker centres as reasonable restaurant standards which are regulated by the National Environment Agency (NEA) that monitors food safety and hygiene. Kim's study found that food activities do not influence significantly on any of the brand equity process based on the locals' perspective. This is because the locals perceived food activities are publicity events instituted to promote local foods to international tourists. Such events are pivotal to boost Singapore's country image as a global culinary destination.

Gastronomic *Kopitiam* and Hawker Culture

A discovery of Singapore's historic ethnic enclaves appeals to both tourists and locals in this city. One, the *kopitiam* culture where *kopi* is the Malay word for *coffee* and *tiam* is a Hokkien word for *shop*. Two, the mouth-watering ethnic food hawker culture. The hawker culture embeds the evolving story of migration and social-cultural diversity.

Kopitiam (coffee shop in local Chinese dialect) or *kedai kopi* in Malay are found all over Singapore, including public housing estates where 85% of the island city's population live. It is not unusual for residents to find their way to *kopitiam* to rest, drink and eat in their walking trails. This archetypal feature of Singapore's daily life has been transformed from its earlier days in neighbourhood streets of small villages serving nibbles, drinks and meals during the colonialist era of immigration to Mod-Sin (Modern Singapore) themed restaurants and stalls serving food rooted in Singapore's traditional and diverse culinary canons. Mod-Sin attests to Singaporeans' enduring self-identity and fondness for the multifariousness of its local foodways.

The *kopitiam* gastronomic trails started mostly at the Singapore River and its banks. Businesses flourished. The waterway was getting polluted as businesses including street hawkers dumped their waste into the river. Kallang Basin and its rivers were not spared from the pollution, with duck and pig farms and shipyards adding to the problem. In March 1969, then Prime Minister Lee Kuan Yew asked for a waterway clean-up plan including curbing further pollution. Spearheaded by PUB and the Public Works Department (PWD), massive work to clean-up the Singapore River and Kallang Basin took place between 1977 and 1987. Close to 5,000 street hawkers, squatters and small industries (for example, charcoal dealers and boat builders) were relocated. Street hawkers were moved to purpose-built hawker centres such as Boat Quay, Chinatown and Empress Place. Several government-built hawker centres in the 1970s were erected to house street hawkers. The government stopped building hawker centres in 1986 after successfully rehousing the street hawkers. Private sector saw a demand for hawker centres in certain areas and started to build them, driving up rentals which, in turn, raised food prices. Calls were made for the government to resume building hawker centres which the government did following the General Election in 2011.

The closure of certain old stalls and kopitiams gives rise to narratives of business expansion, revitalisation and the preservation of those early eateries and coffee shops. Successive family generations now open and manage many outlets all over Singapore like the Indonesian-Malay Sabar Menanti food chain, South Indian Komalas and Bhavan, Hainanaese Yakun, Hans and Jack Place. There are a few inspiring stories to tell. Lim Bee Huat, who started as a nine-year-old *kopi kia* (coffee boy in Hokkien) and is now the *Kopitiam* chain boss. Pang Lim, who started as a kitchen helper when he was thirteen years old is now the boss of *Koufo* (Chinese belief that is good fortune to feast on good food) foodcourt chain.

Kopitiam heritage will be here to stay, attracting budding entrepreneurs such as Lim Bee Huat and Pang Lim to enter the market. One example is a Western chain, Botak Jones. Founded by Bernard "Bernoe" Utchenik in 2003, a Detriot native turned Singaporean in 2009, the food chain is well known for serving generous-portion and hearty *ang moh* hawker food. The long-time food operators and *kopitiam* appeal to its heartland dwellers who travel on foot and/or public transport to eat and drink in their premises, bringing back old memories, nostalgia and pride, largely because the local *kopitiam* is an apparent element of culture, heritage and tradition of Singapore. Despite the fusion of Western business and modern management techniques, the new generation of *kopitiam* chain owners continue to perserve the uniqueness of Singapore's hawker culture and the good old coffee shop traditions. Personalised service with *friendliness and intimacy* are easily forged between the kopitiam staff and customers walking into their outlets such as the Hainanese-themed Killiney *kopitiam* chain with its tagline—*Welcome to the good old days*. On a similar vein, Ya Kun's mission is about preserving its unique and rich heritage and its branding thrives on the belief that—a *good toast binds kinship, friendship and partnership*.

Eunice Tan and Barkathunnisha Abu Bakar cited food-related activities as a key element of the socio-cultural landscape of Singapore. A destination's culinary heritage becomes a primary cultural resource that attracts the locals and tourists, due to its exceptional charm, local identity and sense of place.[10] Tourists and locals alike can indulge in the local cultural heritage by venturing on foot or bicycle through food trails. These trails allow them to not only taste the authentic flavours but also immerse themselves in the vibrant life of the neighbourhood. To enhance their experience, there is a rise in the use of mobile applications that provide food ratings and directions to specific kopitiam outlets and hawker centres. This enables visitors to navigate the culinary landscape with ease. In Singapore's diverse neighbourhoods, a mingling of locals and tourists creates a dynamic culinary tourism scene where unique experiences are formed.

Wu Ting recognises hawker centres as the favourite places for both tourists and locals wanting scrumptious food at affordable prices in Singapore. Hawker centres are known as the nation's Food Paradise.[11] The modern ones are referred to as food courts and these gastronomical destinations offer thousands of food stalls islandwide. A wide range of local food are served to this food community from *Bah Kut Teh* (Pork Bone Tea soup originated from China's Fujian or Hokkien province), Black

[10] Tan and Abu Bakar (2018).

[11] Wu (2021).

Pepper Crab, Chicken Rice, Fried *Kway Teow* (stir-fried Chinese flat rice noodles), *Hokkien Mee* (Prawn noodles from China's Fujian or Hokkien province), *Laksa* (rice noodles cooked in hot soup or curry sauce, a Malaysian dish of Chinese origin), Satay (small pieces of meat grilled on a skewer and served with spiced peanut sauce, an Indonesia and Malaysian food), as well as *Yong Tau Foo* (deep fried meat paste in stuffed bean curd skin accompanied with soup). These gastronomic destinations have much to offer for those dining in with the company of family and friends.

Anyone patronising the food stalls in Chinatown Complex Food Centre at Smith Street after a walkabout in this rich heritage Chinatown is in for a treat. There are a variety of food menu from over 260 food stalls and the local traditional dishes come at cheap prices. The largest of its kind, food lovers can find all the must-try local dishes of Singapore. It is about one minute walk from the nearest Chinatown MRT station.

Old Airport Road Food Centre, which is a one-minute walk from Chinatown MRT station, is one of the oldest hawker centres in the city of Singapore. This food court boosts the highest number of popular food stalls and is home to some of the well-known soya bean curd stalls, the 51 Soya Bean curd and *Lao Ban* Soya Bean curd.

For foodies, the Maxwell Road Hawker Centre is not to be missed out as it offers extremely low-priced Chinese cuisine. Its savoury food has been featured on television channels for its excellent authentic local food. The following food stalls are the ones that are most popular including the China Street Fritters, Hoe Kee Porridge, Marina South Delicious Porridge, Tian Tian Hainanese Chicken Rice and Zhen Zhen Porridge stall. It is a six-minute walk from the Tanjong Pagar MRT station and after a satisfying meal, visitors can visit the beautiful Buddha Tooth Relic Temple and Museum opposite the road and continue exploring the Chinatown area.

Tiong Bahru Market Hawker Centre is one of the most recently renovated food courts and with its latest mechanical exhaust system fitted in its stalls, many comb the stalls after taking about an eight-minute walk from Tiong Bahru Station. It is tipped to be among the cleanest and most spacious hawker centres in Singapore. One can enjoy some of the most popular Singapore food such as *bak kut teh* (Pork Bone Tea soup), *chwee kueh* (steamed rice cake served with preserved radish) chicken rice, herbal soups and *wonton* mee (egg noodles in hot broth with dumplings and leafy vegetables).

Those who are up for a longer walk can take a twenty-one-minute walk from the Lorong Chuan MRT to the Chomp Chomp Food Centre which is open till the wee hours daily. It serves the best-tasting beef, chicken, mutton and pork satays (meat skewers) and delicious local foods such as barbequed chicken wings, carrot cake, fried oyster, grilled stingray, *popiah* (spring roll made from thin flour skin wrapped around meat and chopped vegetables) and prawn noodles.

A three-minute walk from Tanjong Pagar MRT is the two-storey Amoy Street Food Centre built in 1983 and hot-spot for traditional hawkers and modern fusion fare at cheap prices. It is a local culinary institution which had six stalls listed in the Michelin Guide Singapore 2019, and four of them were awarded the Bib Gourmand distinctions. Its wide range of stalls is a revelation for diners, hawking from sliced food soup to curry puffs.

Lau Pa Sat Hawker Centre is a five-minute walk from Raffles Place MRT and accommodates up to 2,500 people at one time. It has striking architecture due to its octagonal shape with its ornamental columns and offers good hawker fare for those hoping to rest their feet and looking for a satisfying meal. After a walkabout in the Central Business District, visitors can try some of the best local delights and international cuisines such as Vietnamese, North Indian and Korean.

There is no doubt the *kopitiams* are a prototypical component of Singaporean culture and this becomes *official* after being included in the UNESCO list of Intangible Cultural Heritage in mid-December 2020.[12] Kids ordering *Milo* dinosaurs (chocolate malt-based beverage) to adults taking away *cai fan* (economic rice) lunches to office is a common scene. The hawker centre provides a venue for uncles and aunties to gossip and share trades and secrets. Old tissue packet *chope* (a Singapore slang) refers to reserving a seat in a place to eat, along with another common method, namely placing of spoons on the cup topsso no one will clear it from the table. *Teh* at kopitiam is made from tea dust or broken tea leaves, using a tea sock to strain it out. Coffee beans in *kopi* (coffee in Malay) are pre-fried in butter or margarine and is an all-time favourite for breakfast, this method is still used till today that gives the *kopi* a creamy and rich taste. Patrons can request their drink stall vendor to serve their *teh* (tea in Malay) or *kopi* (coffee in Malay) *ban shao* (half hot). *Cai fan* or *cai*

[12] Kwek (2021).

png (vegetable rice or simply mixed food and rice) is a permanent feature that locals and tourists may find the food galore too tempting. Most food stalls whether serving *bak kut teh* (Pork Bone Tea soup), chicken rice, or fish ball noodles, offer free refills of soup or the flavourful broth.

This chapter has explored Singapore's status as a gastronomic city-state with its cultural iconic Hawker Culture. Local culinary culture assumes the task of gastrodiplomacy imbued in place branding. Singapore's food restaurants and outlets are most significant in influencing all three dimensions of the brand resonance model—the awareness of Hawker Culture, love of Hawker Culture and loyalty to Hawker Culture. The awareness is linked to the local ingredients and local cooking skills factor. The awareness and love are associated with the education (cognitive aspect) and training (affective aspect) factors. It is obvious to Singaporeans; the hawker centre is the most common food and beverage outlet patronised. Locals just love their street food culture.

Kopitiams are the favourite places for locals (and tourists) wanting scrumptious food at affordable prices in Singapore. Hawker centres are known as the nation's Food Paradise. The modern ones are referred to as food courts and these gastronomical destinations offer thousands of food stalls islandwide. A wide range of local food are served to this food community from *Bah Kut Teh* (Pork Bone Tea soup originated from China's Fujian or Hokkien province), Black Pepper Crab, Chicken Rice, Fried *Kway Teow* (stir-fried Chinese flat rice noodles), *Hokkien Mee* (Prawn noodles from China's Fujian or Hokkien province), *Laksa* (rice noodles cooked in hot soup or curry sauce, a Malaysian dish of Chinese origin), *Satay* (small pieces of meat grilled on a skewer and served with spiced peanut sauce, an Indonesia and Malaysian food), as well as *Yong Tau Foo* (deep fried meat paste in stuffed bean curd skin accompanied with soup). There is no question that the *kopitiams* are an archetypal component of Singaporean culture and this becomes *official* after being included in the UNESCO list of Intangible Cultural Heritage.

Food-related activities are a key element in the socio-cultural landscape of Singapore. A destination's culinary heritage becomes a primary cultural resource that attracts the locals and tourists, due to its exceptional charm, local identity and sense of place. Visitors could go on food trails to get a taste of the local cultural heritage. They may also partake in the colourful life of the local neighbourhoods by touring and eating. There is an increasing number of mobile-driven applications to better enable locals and tourists on these food trails. They get to know the food ratings and directions to reach a particular *kopitiam* outlet or hawker centre. In these Singapore neighbourhoods, locals and tourists mingle, forming their own culinary tourism experiences.

References

Azadi, R., Yousefi, B., & Eydi, H. (2015). The impact of brand country-of-origin image on the formation of brand equity in the sports apparel industry. *Universal Journal of Industrial, 3*(3), 67–73. https://doi.org/10.13189/ujibm.2015.030301

Cannarella, G. (2022, June 8). *Gastro-diplomacy: The science that values food as an identity*. https://www.finedininglovers.com/article/gastro-diplomacy. Accessed on 17 November 2022.

Chapple-Sokol, S. (2013). Culinary diplomacy: Breaking bread to win hearts and minds. *The Hague Journal of Diplomacy, 8*(2), 161–183.

Cheng, X. M. (2017). "City of gastronomy" of UNESCO Creative Cities Network: From international criteria to local practice. *Social Systems Studies*, 55–67.

Clarke, D. (2016). Theorising the role of cultural products in cultural diplomacy from a cultural studies perspective. *International Journal of Cultural Policy, 22*(2), 147–163.

Gordin, V., & Trabskaya, J. (2013). The role of gastronomic brands in tourist destination promotion: The case of St. Petersburg. *Place Branding and Public Diplomacy, 9*(3), 189–201.

Jung, H. J., Lee, Y., Kim, H., & Yang, H. (2014). Impacts of country images on luxury fashion brand: Facilitating with the brand resonance model. *Journal of Fashion Marketing and Management, 18*(2), 187–205.

Keller, K. L. (2001). Building customer-based brand equity. *Marketing Management, 10*(2), 14–19.

Kim, B., Lee, D. C., Chua, B. L., & Han, H. (2022). Country image of gastronomy and branding Hawker Culture: Local consumers' perception. *Tourism Management Perspectives, 44*. https://doi.org/10.1016/j.tmp.2022.101036

Kwek, M. (2021, March 20). *8 Kopitiam secrets shared by Hawker Centre aunties & uncles that you should know besides tissue Chope-ing*. https://thesmartlocal.com/read/kopitiam-secrets/. Accessed on 12 September 2021.

Pappu, R., Quester, P. G., & Cooksey, R. W. (2007). Country image and consumer-based brand equity: Relationships and implications for international marketing. *Journal of International Business Studies, 38*(5), 726–745.

Rockower, P. S. (2012). Recipes for gastrodiplomacy. *Place Branding and Public Diplomacy, 8*(3), 235–246.

Tan, E., & Abu Bakar, B. (2018). Ethnic reunions in tourism: The route to roots. In *Asian cultures and contemporary tourism* (pp. 21–39). Springer.

The Economist. (2022, February 21). Thailand's gastro-diplomacy. https://www.economist.com/asia/2002/02/21/thailands-gastro-diplomacy. Accessed on 18 December 2022.

Wu, T. (2021). *7 Best Hawker Centres you should visit in Singapore*. https://bestinsingapore.com/hawker-centres/. Accessed on 10 September 2021.

CHAPTER 6

Natural Heritage

Singapore's remarkable evolution into a garden city can be attributed to years of unwavering commitment and continuous efforts to embrace greenery. Initially, the focus was on providing ample shade and accessible green spaces for its residents. Over time, the government, in collaboration with the community, has nurtured and expanded these greening initiatives, embracing biophilic designs, restoring habitats and creating interconnected park networks. The result is a harmonious blend of nature and urban living, where the benefits of green spaces are integrated seamlessly into Singapore's landscape, enhancing the well-being and quality of life for all.

Singapore's early park development programme was led by Parks and Recreation Department. Since its establishment in 1976, the number of parks has increased from 13 to 330 between the period of 1975 to March 2014. The area of green spaces and parks grew to 9,707 hectares from a mere 879 hectares in the same period.[1] The Parks and Recreation Department was reconstituted as National Parks Board (NParks) in July 1996. NParks extends Singapore's natural capital and restore nature amid Singapore's urban landscape, which entails a network of green space, parks and

[1] Public Works Department (1975).

Melissa Liow Li Sa and Sam Choon-Yin, *Sustainable Urban Development in Singapore*, Sustainable Development Goals Series,
https://doi.org/10.1007/978-981-99-5451-3_6

park connectors. While past conservation efforts had centred on planetary aspects, fresh prominence has swung to include marine environment, pictured by the turnaround of development works on a distinctive intertidal shore called Chek Jawa, and the re-designation of Labrador Park as a nature reserve and Sungei Buloh as Wetland Reserve. In tandem, Singapore roll-out a ten-year *City in a Garden* plan to heighten the city's natural infrastructure, and to further involve and engage the public members.

This chapter discusses Singapore's Parks, Trails and Gardens Nature Reserves, Marshes and Wetlands and Islands, which are the nation's exemplars in imbuing the green infrastructure concept. Rural tourism, marine national park and heritage conservation, mangrove habitats and biodiversity protection, as well as coastal protection and restoration are also discussed to support Singapore's Green Plan 2030 and Sustainable Development Goals (SDGs).

Green Infrastructure

Green infrastructure encompasses a range of nature-based solutions that contribute to sustainable urban development. It involves the strategic integration of natural elements, such as parks, green spaces, water bodies and trees, into the built environment. In Table 6.1, we explore the multifaceted nature of the term "green infrastructure" and highlight its various meanings across different contexts.

Mark Benedict and Edward McMohan's definition is appropriate for Singapore's urban environment, as it encompasses metropolitan, bioregional and site scales. Open spaces are essential for promoting smart growth on a metropolitan scale, while on a bioregional scale, they support and preserve vital ecosystem functions through the implementation of green infrastructure. At the site scale, these open spaces facilitate the integration of business and social activities. At the lower scale, green infrastructure efforts include green roofs, infiltration planters, porous pavements, rain gardens, rainwater harvesting for non-potable uses, trees and tree boxes. The green infrastructure planning and design comprises three scales: state-wide (national forests connect the state through green corridors), community (existing parks are connected with greenways)

Table 6.1 Meanings of green infrastructure

Different meanings of Green infrastructure	*Citations*
Environmentally friendly engineered structures with green features that are either facilities or substructures, comprising a preservation system that focuses on the social and ecological influences of sprawl and the increased open land fragmentation pace. Additionally, it is an interconnected network to preserve natural ecosystem functions and values, maintenance of clean water and air and to benefit people and wildlife	Benedict and McMahon (2001)
Natural and managed green open spaces in rural and urban settings that offer strategic linkages to a myriad of amenities for people to enjoy	Davies et al. (2006)
Farmlands, forests, mountains, parks and playgrounds and stream valleys are constructed as natural open areas. Urban forestry and parks can help to maintain the air temperature and cleanliness, as well as reduce the energy consumption and costs	Jerke (2008)
Varying urban green infrastructure and landscapes that encompass allotments, green corridors, parks, private gardens, public green spaces, roof and vertical greening, street trees and urban forests	Cameron et al. (2012)
A planning instrument that contributes to socioeconomic benefits through competitive, inclusive, sustainable and resilient urban spaces. Blue and green ecosystems are strategically planned networks of semi- and full-natural spaces that deliver a comprehensive array of ecosystem services of different degrees	Monteiro et al. (2020)

Sources Benedict and McMahon (2001), Cameron et al. (2012), Davies et al. (2006), Jerke (2008), and Monteiro et al. (2020)

and individual parcels (businesses and homes are located around green spaces).[2]

[2] Benedict and McMahon (2001) posit seven principles for green infrastructure efforts:

Principle 1: Sharing similar strategies pertaining to planning and protecting interconnected green space systems.

Principle 2: Demarcate in advance and conserve crucial ecological sites and linkages as restoration is more costly than protecting the prevailing landscapes.

Principle 3: Linkages of public-private sector and non-governmental organisations (NGOs) are important in preserving critical wildlife biodiversity, services, and ecological processes.

In Singapore, landscape linkages (large protected natural areas) are connections to green infrastructure networks (natural preserved areas and existing parks) to enable green infrastructure systems to operate. Other than functioning as corridors connecting landscapes and ecosystems, adequate space is provided for native flora and fauna to thrive. Landscape linkages offer additional space for recreational purposes and protection of historic places.[3] Indeed, Singapore has an integrated system of parks and parkways, providing a comprehensive network of urban open spaces than just individual, disconnected parks. Urban open spaces such as Singapore's 4.3-hectare *Padang* marks many iconic National Day Parade celebrations.[4] The *Padang* is bordered by Connaught Drive, St Andrew's Road and Stamford Road and it is linked to the Esplanade Park. The Esplanade Park is an important frontage connecting the Supreme Court, *Padang* and City Hall, a heritage park built in 1943. This particular area of the city has flourished with vibrant human interactions centred around these significant historical symbols. The surrounding environment serves as a testament to Singapore's commitment to open spaces and natural systems, which are crucial considerations in the development of the city's urban landscape.

In the following sections, we will explore notable examples of how modern-day Singapore is dedicated to preserving its invaluable natural heritage.

Principle 4: Systems of green infrastructure must link the wilderness, rural, suburban, and urban surroundings, while incorporating these green space functions and components across numerous jurisdictions at varying degrees.

Principle 5: Fundamentally land-use planning practices and theories and sound science, green infrastructure overlaps with other fields such as conservation biology, geographic analysis, information systems, landscape ecology, regional and urban planning.

Principle 6: An annual budget as form of public investment by the federal and state agencies dedicated to providing resources for green infrastructure planning and management activities.

Principle 7: Require involvement of various stakeholders or organisation in establishing strategic alliances to enable green infrastructure efforts to bear fruition.

[3] Pertaining to connectivity amidst green infrastructure systems, Marshall (2004) describes *connectivity* as the sum of routes or links from and to a specific destination, and *permeability* as capacity of these connections to ferry vehicles or persons. Connections form corridors that link one to another, by open spaces, pedestrian ways and streets that are positioned fragmentedly across the city.

[4] The *Padang* has become a vantage point to visitors as well as a historical site that occupies a special place for its local people. Sports and recreational activities such as line dancing, cricket, hockey, football and rugby are hosted on the *Padang*. Tan (2022).

Nature Parks, Reservoir Parks, Ridge Parks

Singapore has more than 300 parks occupying a land area of more than 550 hectares, as well as mangroves, nature reserves, coral reefs and tidal flats. The small country is also home to a wide range of flora and fauna. Forty-seven per cent of the land area is covered in green spaces. By 2020, a typical household will not take more than 10 minutes to walk to a park.

Singapore's largest nature parks is none other than the Chestnut Nature Park. Occupying a total area of 81 hectares, Chestnut Nature Park is one of the networks of parks that partially surrounds the Central Catchment Nature Reserve. The park is popular among families as it offers a relaxing walk due to its relatively flat hiking trails with occasional slopes. The park is suitable for mountain biking and has 11 trails with varied difficulty levels. The north of the nature park contains a 3.5-kilometre hiking trail that features the Chestnut Observation Tower and its south is a 2.1-kilometre route that has rockier paths, steeper inclines and tougher terrain. Visitors normally spend about 2 hours and can get to see rare wildlife as well as hidden lakes and streams along the park trails in this 81-hectare land.

The Bukit Batok Nature Park, formerly a granite quarry site, offers a picturesque setting with its stunning quarry pond. Visitors typically spend an average of 1.5 hours exploring the park. In the past, there was a Japanese World War II memorial at the site, but it has since been destroyed, leaving behind only two short pillars and 120 steps. The area is now occupied by the Mediacorp Transmission Centre, Singapore's state-owned media conglomerate. The park caters to both adventurous hikers seeking off-the-beaten-path forest trails and leisurely walkers who can enjoy the serene views of the park's pristine lakes, remnants of its former quarry.

The Dairy Farm Nature Park used to be a cow shed. Via the Dairy Farm Loop, the nature park offers a challenging trek leading to Bukit Timah Hill. Nestled just a short fifteen-minute walk away from Dairy Farm Road Carpark A, awaits a true treasure—a peaceful quarry unlike any other in Singapore. Unlike the typical quarries that are transformed into lakes, this exceptional quarry remains filled with earth, exuding a unique charm and serenity.

Fort Canning Park, with its nine enchanting gardens, has become a beloved destination for both wedding photography enthusiasts and nature lovers alike. The 2-kilometre history-themed trail brings history buffs to the First Botanic Garden back in 1822 and the Sang Nila Utama Garden which was named after the first ancient king and inspired by the fourteenth-century Javanese gardens with freshwater features and Javanese split gates.

HortPark is a zen-inspired gardening centre and home to ten beautiful, themed gardens. The more well-known Therapeutic Garden has design elements that are targeted at the elderly. The park has its own Bee Trail, one of the city's buzzing pollinators and the 30-minute walk will take one through three distinctive bee habitats with over 20 bee species.

MacRitchie Reservoir Park is the oldest and most well-known nature park in Singapore. Amid its rainforest, there are both ambitious hikes and easy boardwalk treks, 3–11 kilometres in length comprising the MacRitchie Trails surrounding the reservoir. Key features include the suspension bridge which is situated 25-metre above the forest floor as well as the TreeTop Walk, linking the two highest points in the reserve. Visitors can come close to an impressive variety of wildlife such as lemurs, long-tailed macaque monkeys, pangolins and tree frogs. The park offers beautiful tropical scenery, peaceful and secluded paths ideal for trekking in Singapore with a clear view (Instagram worthy) of the forest canopy. Other attractions include the 11-kilometre round-trip hike around the park and the eight-storey Jelutong Tower.

Mount Faber Park offers spectacular city scenes. One of the oldest parks in Singapore, the 0.8-kilometre trail, which takes about 15 minutes to complete from the starting point at Mount Faber Carpark, brings visitors to the tallest pedestrian bridge—the Henderson Waves. The park offers shade from its dense rainforests and one can spot several bird species along the walking trails.

Pasir Ris Park has its own Mangrove Boardwalk, offering visitors an up-close encounter with mudskippers, monitor lizards and tree-climbing crabs. The park has a stunning coastline where adults can fire up the grill for a barbeque outing and kids to burn energy in its maze gardens and playgrounds. Bird watchers can head to the three-storey viewing tower and a perfect vantage point to spot the White Collared Kingfisher.

The Punggol Waterway Park has something in store for everyone of all ages. Visitors can admire the scenic landscapes from the Kelong Bridge and Rope Bridge. The park has four unique themes suitable for inline skating activities and cycling. The Green Gallery offers its visitors a peaceful trail that runs along its natural terrain. The Heritage Zone allows the visitors to take a walk down the memory lane amid the beautiful greenery along the old Punggol Road. The Nature Cove has a relaxing lawn area for families to spend their lazy weekends with a quaint view of the Waterway. Kids can play at its Recreation Zone, featuring sand play and water play.

Rower's Bay Park derives its name from its reservoir, which is a popular destination for kayaking enthusiasts. For those who prefer to stay on dry land, the park offers the opportunity to explore various plant species that provide shelter and food for the local wildlife. Visitors can also enjoy the park's boardwalk, which allows them to discover the surrounding wetlands without getting their feet dirty. Rower's Bay Park is part of the initial phase of the planned 150-kilometre Round Island Route (RIR), a project aimed at connecting parks across the entire island. The completion of the route in 2035 will provide cyclists with a safe and comfortable environment to cycle around the island.

The Tampines Eco Green is linked through park connectors from the Tampines Biking Trail and Sun Plaza Park. Its eco-friendly route permits visitors to walk through its secondary forest, open grasslands and freshwater wetlands, and get close to the flora and fauna surrounding the route. Birdwatchers may spot the famous Baya Weavers. Visitors can leisurely walk within the life sanctuary or sit on one of the rustic wooden benches.

The Windsor Nature Park is located near Upper Thomson. Despite its Downtown Abbeyesque name, it is a wetland that is befitting of Singapore's Garden City concept. The park has a 150-metre-long elevated walkway and four exclusive trails over a 75-hectare land. Its Hanguana, Squirrel and Drongo trails are mostly wheelchair friendly. Its Drongo trail offers a sub-canopy walk where it's park visitors can walk through the forest and admire its beautiful flora. Other park features include its marsh habitat and freshwater streams.

Upper Peirce Reservoir Park provides a leisurely trekking experience for individuals who enjoy easy-going trails along the lakeside. This serene park is inhabited by sizable populations of long-tailed macaques, necessitating caution with food items to prevent them from being snatched away. Visitors have the option to extend their hike to Lower Pierce Reservoir Park enveloped by dense forests that offer a serene and refreshing atmosphere.

The Southern Ridges offers a 10-kilometre scenic path linking five parks from Labrador Nature Reserve, Kent Ridge Park, Hort Park, Telok Blangah Hill Park and Mount Faber Park. One normally spends four hours from the Marang Trail Entrance which is a 6-minute walk from Harbourfront MRT Circle Line or Northeast Line. One of the pinnacles is the Marang Trail which is the harshest part of the path with stairs ascending to a height of a 24-storey building. Other features include the Singing Forest where one can spot colourful birds, and the elevated Forest Walk. For those seeking a leisurely walk, an enjoyable option is to stroll along the gently sloping paths of Henderson Waves and Faber Walk. There are scenic spots that are perfect for nature lovers, bird watchers, history buffs and everyone in between and it is a terrific getaway from the city's hustle and bustle.

Botanic Gardens and Lake Gardens

Singapore Botanic Gardens, established in 1859, holds the distinction of being the first tropical garden to be recognised as a UNESCO World Heritage Site. This expansive park offers a serene environment for leisurely walks and picnics, with most of its grounds situated on flat terrain. In 1875, the garden came under government control and became linked to the Royal Botanic Gardens in Kew, London. Its visionary director, Henry N. Ridley, championed rubber as a commercial crop prior to the rise of the automotive industry.

A major attraction within the Botanic Gardens is the National Orchid Garden, which boasts an impressive collection of over 2,000 orchid hybrids and 1,000 species. Notably, it is home to Singapore's national flower, Vanda Miss Joaquim. The park also features the Learning Forest, a century-old rainforest integrated with a six-hectare extension, creating an expansive habitat for diverse flora and fauna.

Covering a total distance of 6.4 kilometres, the Botanic Gardens offers multiple entry points, with the Bukit Timah Gate conveniently located near the Botanic Gardens MRT station on the Circle Line or Downtown Line. In 2021, an eight-hectare extension was added to the already impressive gardens, including the House, which now hosts Asia's largest collection of high-elevation montane orchids. Visitors will be captivated by the enchanting COMO Adventure Grove, a magnificent play area surrounded by the beauty of nature.

Jurong Lake Gardens is a gigantic 90-hectare park and the country's biggest manufactured floating wetlands at Jurong Lake. The 3.3 kilometres path takes a visitor about one-and-a-half hours to cover the gardens area from its starting point which is 2 minutes walk away from Lakeside

MRT. Visitors can snap photos at its lush Grasslands or *unravel their inner child side* at Forest Ramble. Other than snapping Instagrammable photos that capture the beautiful view of the lake, visitors can look forward to its Japanese Garden and its beautiful water lily pond. The park is home to picturesque waterside trails including the Freshwater Swamp Forest, the swampland at Alstonia Island, the Neram Streams, the Clusia Cove water play area and when strolling down the Rasau Walk boardwalk.

Coast-to-Coast Trails

The Coast-to-Coast (C2C) Trail is Singapore's longest hiking trail, spanning over 36 kilometres and interconnected through various park connectors. The trail offers trekkers a unique experience as they traverse parks, new neighbourhoods and natural landscapes. The journey begins at Jurong Lake Gardens in West Singapore and extends all the way to the Northeast Coney Island Park. Along this remarkable trail, hikers will have the opportunity to explore some of Singapore's finest nature reserves and parks, including the iconic Singapore Botanic Gardens, the picturesque Bukit Batok Nature Park and the serene MacRitchie Reservoir Bishan-Ang Mo Kio Park. Throughout the trail, trekkers will also encounter diverse environments, such as ancient cemeteries, lush forests and enchanting quarry lakes, adding to the richness of the C2C experience.

Rail Corridor route offers a coast-to-coast trail that stretches from West to East. It starts with the Tanjong Pagar Railway Station and was a primary railway track to transport goods between Singapore and Malaysia until it was discontinued in 2011. Many hikers flock to the iconic 1930s-era Bukit Timah Railway Station, which has been sensitively restored to retain its old-school charm. There are routes where hikers can gain rainforest experience with several of its lookout decks hovering the nature's beauty. The Rail Corridor spans a total distance of 24 kilometres and the starting point at Tanjong Pagar Railway Station is about a 14-minute walk from the Tanjong Pagar MRT East West Line. Occasionally known as the Green Corridor, the Rail Corridor appeals to many trekking enthusiasts. Visitors can experience the untouched beauty of the Green Corridor, where they can immerse themselves in the vibrant ecosystem of marshes, streams and grasslands teeming with life. This natural haven provides a unique chance to get up close to the wonders of nature. Moreover, the Green Corridor serves as a vital link, connecting various parks and natural reserves. Notable destinations along this corridor include the captivating Kranji Wetlands and the renowned Bukit Timah Nature Reserve. Together, these interconnected green spaces offer visitors an extraordinary journey through Singapore's diverse natural landscapes.

Nature Reserves, Marshes and Wetlands

Singapore's rich natural heritage, including its nature reserves and diverse flora, has been meticulously documented for over a century, showcasing the nation's ongoing efforts in botanical collection. Through these records, new species have been constantly discovered and added to the collection, with an impressive count of 843 forest angiosperm species identified and recorded in the Bukit Timah Nature Reserve alone. Significant exploration and collection activities have been conducted in the Central Catchment Nature Reserves, particularly in previously undersampled areas, leading to remarkable findings. In fact, some species that were once thought to be extinct have been rediscovered, highlighting the importance of conservation efforts and the continuous surprises that nature has in store for us. Richard Corlett, a professor at the Chinese Academy of Sciences has written extensively on the nation's native. Deforestation from the nineteenth century, Corlett told us, had reduced forest areas to isolated plots in a matrix of grasslands.[5] Almost the entirety of the primary forest areas outside the nature reserves were cleared. However, protection to allow the growth of secondary forests within the reserves has continued, and whatever is left of the primary forests that host distinct species.

Bukit Timah Nature Reserve is Singapore's tallest hill at 163 metres. It encompasses a distance of 3.6 kilometres, which takes an average of 1.5 hours to climb to the hill. The nature reserve is one of the most diverse and richest ecosystems in the world. There are four walking trails for nature lovers, walkers and mountain bikers. The green or route 3 is a winding forest path that leads to caves. There are unpaved trails and a steep paved path which take hikers straight up to the peak. The adjoining Hindhede Nature Park offers visitors a breathtaking view of the quarry lake.

[5] See, for example, Corlett (1995, 1997).

Labrador Nature Reserve, a biodiverse park brimming with history, presents a captivating destination for nature enthusiasts. Designated as a nature reserve in 2002, the park provides scenic walking trails that lead to Singapore's colonial era military defence positions. Within the reserve, visitors can explore the mainland's only rocky sea-cliff, which offers a series of coastal walks. The area is also home to World War II tunnels and hosts over 70 bird species. Retaining remnants of a fort that safeguarded Keppel Harbour since 1878, Labrador Nature Reserve proudly displays war relics like a machine gun post. To further enhance the experience, the 10-kilometres Southern Ridges trail connects Labrador Nature Reserve with other parks, including Mount Faber. Situated a convenient 2.1 kilometres away from the Labrador MRT Circle Line station, the reserve beckons visitors to witness its vibrant ecosystem. Spotting resident squirrels darting up trees and hearing the melodic tunes of black-naped orioles and oriental magpie robins is a common delight. Explorers are encouraged to capture the beauty with their cameras as they embark on the Labrador Nature Reserve boardwalk. This elevated pathway spans across marshes, cliffsides and secondary forests, offering stunning views of the picturesque coastline. The park also treats visitors to an exceptional panoramic vista of the South China Sea, creating a truly immersive and memorable experience.

Kranji Marshes is the largest freshwater wetland of 57 hectares and is a paradise for birdwatching. Home to 170 bird species, explorers may get to spot the threatened bird species like the Grey-headed Fish Eagles and Changeable Hawk Eagle in the woodland areas. Guided NParks tour is available for visitors. The recommended time spent in the 2-kilometre marshes is 1.5 hours and one can hop onto a bus from Kranji MRT to reach its starting point at Kranji Gate.

Sungei Buloh (Bamboo River in Malay) Wetland Reserve is another hiking trail in Kranji, Singapore. *Sungei Buloh* Wetland Reserve was officially opened in 1993 and gazetted as a nature reserve in 2002. The reserve hosts a variety of species of mangroves, plants and animals. One of the iconic attractions is its famous Kingfisher Pod where people can capture the sunrise golden rays and its name was inspired due to many bamboos that once thrived in the reserve. Its 4.3-kilometre distance and seaside boardwalks enable the visitors to catch the nearby views of its neighbouring country, Malaysia. One can spot the wildlife among the mangroves from crocodiles to kingfishers, mudskippers and snakes. When exploring the park's trails, especially during the migratory season, trekkers are treated to the sight of flocks of shorebirds, including plovers and sandpipers.

Islands

Given its small size, Singapore faces the challenge of developing a long-term plan to cater to the needs of a growing population, while balancing tourism, recreation, housing, transportation infrastructure and expanding industries and commerce. One potential approach is integration of offshore islands, offering opportunities to optimise land use and meet the diverse needs of the population. Indeed, one of the largest reclamation projects carried out was Jurong Island, where seven small islands—Pulau Ayer Chawan, *Pulau* Merbau, *Pulau* Merlimau, *Pulau* Pesek, *Pulau* Pesek Kechil, *Pulau* Sakra, *Pulau* Seraya, collectively known as the Ayer Islands—were merged to house the petrochemical sector and the world's third-largest oil-refining centre.

According to Brice Laurent and his colleagues, the islands of Singapore provides an ideal setting for combining business experimentation with the concept of island imaginaries.[6] Island imaginaries refer to imaginaries of exceptions. The islands are territories of exceptions and places like no others (exceptional territories). Singapore employs these two modalities of exception in its *Smart Nation* vision in a programme to promote business experimentation. Jurong Island and *Pulau* Bukom are dedicated to the petrochemical industry. *Pulau* Ayer Merbau hosts the nation's first petrochemical complex. *Pulau* Seraya has a power station. *Pulau* Pesek Kechil is the holding station for imported pigs in the mid-1980s. *Pulau* Sakra is designated for future industrial uses.[7] *Pulau* Semakau stands as a remarkable testament to Singapore's innovative waste management practices. Serving as the nation's sole offshore landfill, it employs an advanced incineration process that reduces waste volume by an impressive 90%.[8]

But reclamation is not without cost. It has been estimated that 60% of the island city's reefs have been lost to reclamation.[9] Lionel Ng and Chou Loke-Ming advocate reef restoration as necessary to maintain the

[6] Laurent et al (2021).

[7] See Lee (2000).

[8] See Connolly and Muzaini (2021).

[9] See Lai et al (2015) and Tay et al. (2018).

resiliency of the ecosystems. Strategies include the development of coral nurseries, substrate modification to increase coral survivorship, planting of artificial reefs for colonising reef biota and transplantation for coral cover in sullied areas.[10] These efforts are productive towards restoring ecosystem services. By populating the reef diversity, it can mitigate climate change, enhance visitors' education of the reef colonies, and further legitimise coral protection. The commitment and cooperation of civil society, policymakers and scientists are then critical to ensure the reefs remained a part of the Singapore islands' landscape.[11]

Other islands are transformed for different purposes. Sentosa Island was allocated by the Singapore Government to compensate for the loss of space near Changi in the early 1970s. In the past, the island was known as *Pulau Belakang Mati* (the Island behind Death in Malay)—the island at the outset was primarily populated by Bugis, Malay and Chinese villagers. A mysterious plague in the 1840s annihilated many of the villagers, and then functioned in the nineteenth century, as a British military base before developing into a *tropical island resort and South Sea Island paradise* in the 1970s.[12] The island was renamed Sentosa (which means peace and tranquillity in Malay).[13] The Sentosa Island has ever since hosted a wide range of fun and entertainment attractions for adults and kids while promoting walking, trekking, hiking and cycling for the old and young.

For many, walking around the Sentosa Island exposes visitors to hair-raising iconic attractions from soaking oneself in white-water rafting to hopping on an indoor thrill ride. Families can enjoy a train trip through Sesame Street and take in the aerial views of Jurassic Park on a canopy flyer. For water enthusiasts, Adventure Cove Waterpark beckons with its high-speed water slides, hydro-magnetic coaster and exhilarating water tornado rides. Adrenaline seekers can venture to Hackett Sentosa for bungee jumping, vertical skywalks, giant swings, free falls, and sky bridge crossings. Sentosa Luge and Skyride offers an exciting blend of tunnel

[10] Ng and Chou (2018).

[11] Creighton Connolly and Hamzah Muzaini also noted that as islands were redesignated to serve the developing nation's metabolic and economic needs, the islands' Indigenous communities have been displaced, and the islands' ecologies severely impacted. Connolly and Muzaini (2021).

[12] Turnbull (1972), Campbell (1967), Ho (2015), and *The Straits Times* (1969, January 23).

[13] See *The Straits Times* (1972, March 24).

exploration, downhill slopes and treetop rides, allowing visitors to indulge in the thrill of the tracks and stunning views from the Skyride. Access to Sentosa Island is convenient via a footbridge or the Sentosa Express, providing a seamless connection between the mainland and the island. Aside from the exhilarating attractions, Sentosa Island boasts a vibrant atmosphere with high-end restaurants, shopping malls, spa centres and relaxation spots like Tanjong Beach Club. Accommodation options range from the iconic Hard Rock Café to luxurious hotels like Hotel Capella and Equaris Hotel.

The following islands also exemplify the walkability of Singapore:

Pulau Ubin is a reminder of village-style of living in old Singapore and is deliberately kept in its rustic and natural state. The island is visited by Singaporeans in search of a quiet getaway, a place to rent bicycles to discover the exceptional variety of plants and animals. The island houses the well-known Fo Shan Ting Da Bo Gong temple.

St John's Island is also known as *Pulau Sakijang Bendera*. Tipped to be one of the most picturesque southern islands of Singapore, the island is a popular destination for honeymoon couples and adventure lovers, and the place to go for jungle biking, strolling, beach picnicking and swimming in its blue seas and lagoons.

Lazarus Island offers an idyllic escape just a short distance from Singapore city. It presents a plethora of activities for day excursions, including sunbathing, cycling and swimming. Private ferries can be hired from the Marina South Pier, and some visitors take the opportunity to explore its neighbouring island, St. John's Island. A leisurely stroll on Lazarus Island allows you to soak in the breath-taking scenery of pristine white sandy beaches, clear azure waters and lush green palm plantations that envelop this 47-hectare paradise.

Kusu Island holds a special place in the hearts of spiritual tourists, drawing both Taoist and Muslim devotees who consider the island sacred and holy. Its name, "Kusu", meaning tortoise in Chinese, is inspired by its stunning coral reefs, diverse marine life, and the striking verdant green cliffs that resemble the shape of a tortoise. Upon reaching the island, visitors embarking on a trek to its hilltop will discover three revered shrines: Nanek Ghalib, Puteri Fatimah and Syed Abdul Rahman. These sacred sites add to the spiritual significance and cultural richness of Kusu Island.

Sister's Islands, located approximately 35 minutes away from Singapore city, offer a plethora of exciting activities such as guided walks, snorkelling and dive trails. The islands, namely Subar Darat and Subar Laut, are separated by a deep canal and boast expansive, picturesque beaches that are perfect for water sports and adventurous pursuits. Exploring the marine parks surrounding the islands reveals a vibrant underwater world teeming with fascinating marine life, including octopuses, corals and majestic giant clams.

Walkability and Sustaining the Natural Heritage

Since its reconstitution in July 1996, NParks of Singapore has demonstrated a strong commitment to conserving nature and biodiversity, fostering building capabilities in the organisation and industry, establishing Singapore's reputation and position as a *City in Nature*. The *City in Nature* aims to enhance Singapore's green, liveable and walkable environment. The measures include raising (1) the park provision ratio of 0.83–0.80 hectare/1,000 population (2) 143–200 hectares of skyrise greenery (3) 370–500 kilometres of park connectors (4) 170–300 kilometres of nature ways and (5) 93–100% of households within a 10-minute walk to a park (6) 1,700–3,000 community-in-bloom gardens and (7) 338,800–1,000,000 more trees planted. Table 6.2 outlines five sustainability strategies that underpin the upcoming plans to promote walkability.

Continued efforts are being made to integrate nature into Singapore's urban environment. In April 2020, the OneMillionTrees Movement was initiated in Singapore, aiming to enhance the city-state's natural surroundings by planting an additional one million trees by 2030. Singaporean families with new-borns are offered a 3-year time frame from their birth date to register for a tree planting opportunity under Family Trees; empowering families to produce a greener, more sustainable Singapore collectively, while leaving an enduring legacy for their children.[14] After two years of the programme, more than 540,000 trees have already been planted. Despite the city's need for space and development, Singapore has remained steadfast in preserving its greenery. Trees serve numerous purposes, including providing shade to lower ambient temperatures, enhancing air quality, serving as habitats for diverse forms of wildlife and storing carbon to combat the effects of climate change.

[14] As part of Singapore Together, this FamilyTrees drive encourages Singaporeans and families to take accountability of their sustainable future.

Table 6.2 Five sustainability strategies in developing *Singapore's City in Nature*

Five sustainability strategies	*Aims*	*Present examples*
Sustaining our City in Nature Designing the best viable living environment through the merits of populating greenery, nature conservation and recreation	Extending and Conserving Natural Capital	Between 2020 and 2021, the addition of two new nature parks, namely the Rifle Range Nature Park, which acts as a buffer for Singapore's Bukit Timah Nature Reserve, and the Khatib Bongsu Nature Park, a thriving mangrove and mudflat habitat, expanded the network of nature parks. This expansion aims to preserve and enhance natural carbon sinks throughout the island while also promoting walkability
	Bolstering Connectivity in between Green Spaces	A vast network of 370 kilometres of park connectors has been established, which is equivalent to seven times the length of Singapore from west to east. The eastern section of Stage 1 of the Round Island Route was opened to the public in January 2022, featuring a 75-kilometre trail stretching from Berlayer Creek to Rower's Bay. In addition, two new Coast-to-Coast (C2C) trails will be introduced alongside the existing Central Trail: a 62-kilometre Southern Trail spanning from Tuas to Changi, and a 25-kilometre Northern Trail connecting Sungei Buloh Wetland to Khatib Bongsu. These expansions in recreational green spaces across the island will enhance walkability opportunities for visitors to enjoy An Ecological Profiling Exercise (EPE) was initiated in 2021, and as a part of the EPE, Clementi Nature Corridor was recognised as a significant ecological connector between the Southern Ridges and the Bukit Timah Nature Reserve. NParks worked and completed 170 kilometres of Nature Ways, which mocked the multi-layer formation of a natural forest to increase the linkage between green spaces for fauna such as butterflies and birds, and also beautify Singapore city, providing shade, cool surroundings and improving air quality to drive walkability for people from all walks of life and ages

Five sustainability strategies	*Aims*	*Present examples*
	Intensifying Nature in Parks and Gardens	Approximately 40% of Singapore is covered by gardens, nature reserves, parks, roadside greenery, skyrise greenery and empty state lands. NParks has set a goal to redevelop and develop more than 300 hectares of park spaces by 2026, with a focus on creating natural landscapes and lush vegetation. This expansion aims to provide visitors with many opportunities to engage in walking activities within these natural environments, leading to various health and well-being benefits In order to preserve their natural state, waterbodies such as those found in Jurong Lake Gardens and Bishan-Ang Mo Kio Park were carefully naturalised. Additionally, efforts were made to restore mangroves in Kranji Coastal Nature Park. Mangroves play a crucial role as coastal protection measures, working in conjunction with engineering solutions to mitigate the effects of rising sea levels and intense storms. By preventing erosion and minimising the risk of inland flooding, these restored mangroves serve as a valuable natural resource for safeguarding our coastal areas Parks and gardens now offer therapeutic landscapes that cater to the needs of seniors and young children, providing an opportunity for engaging walkability activities. These landscapes are designed with natural materials and biophilic elements, creating a soothing environment. Children can enjoy child-directed free play, promoting their development and exploration. Additionally, individuals with mood disorders can benefit from the mood-enhancing effects of these landscapes, while those diagnosed with dementia can find comfort and support in their surroundings. These therapeutic landscapes aim to improve the well-being and enjoyment of both seniors and young children through the integration of nature and thoughtful design In 2021, NParks extended the Species Recovery Programme with the aim of enhancing the native flora and fauna of Singapore. The number of targeted species increased from 94 to 120, and it is projected to further increase to 160 species by 2030. Through ongoing efforts in species recovery and habitat restoration, visitors can now encounter once-rare animals such as the Sunda Slow Loris (*Nycticebus coucang*) and plants like the Singapore Ginger (*Zingiber singapurense*) during their walking and cycling trails. These initiatives have contributed to the preservation and revival of diverse species, enriching the natural experiences of visitors in Singapore's outdoor spaces

(continued)

Table 6.2 (continued)

Five sustainability strategies	*Aims*	*Present examples*
	Regenerating Nature amid the Urban Landscape	NParks has been actively revitalising landscape plantings at Mass Rapid Transit (MRT) stations and implementing skyrise greenery initiatives. This has led to an increase in walkability as people enjoy the vibrant and visually appealing vertical green walls, rooftop gardens and green roofs. Additionally, the strategic planting of trees in hotter areas of the island has further enhanced walkability by providing cool and shaded paths. An excellent example of this is Jurong Island, which currently boasts 90,000 trees and is projected to see that number increase to 260,000 trees by 2030. These efforts contribute to a more pleasant and comfortable walking experience, encouraging individuals to explore and engage with nature in urban environments
Nurturing Communities Fortifying Singapore's biodiversity and greenery through partnerships with the community	Support Community Stewardship	Community involvement was integral to the success of Singapore's City in Nature vision. NParks reached out to the community through digital platforms such as organising large-scale virtual concerts, hybrid festivals and programmes during the COVID-19 pandemic
		Starting from November 2021 and continuing into 2022, NParks has introduced a new series of workshops and webinars known as "City in Nature Conversations". The primary objective of these engaging sessions is to promote the City in Nature vision to the public and increase awareness about it. Through these interactive workshops and webinars, NParks aims to foster a deeper understanding of the importance of nature in urban settings and encourage active participation from individuals. By creating a platform for dialogue and learning, NParks seeks to inspire people to embrace and contribute to the City in Nature initiative, making Singapore a greener and more sustainable city

Five sustainability strategies	*Aims*	*Present examples*
		The Community in Bloom (CIB) gardening movement, spanning across the nation, unites residents to foster community stewardship and nurture the spirit of Singapore's City in Nature. NParks, together with its partners, actively promotes walkability by leading and engaging in activities within the CIB gardens. The ambitious target is to increase the number of CIB gardens from over 1,700 to 3,000 by 2030. In June 2020, NParks launched the Gardening with Edibles initiative, which involved distributing free seed packs containing a variety of fruited and leafy vegetables to nearly half a million households in Singapore. Additionally, 400 thousand seed packets were distributed to the public as part of the Singapore Garden Festival Horticulture Show. To support the establishment and operation of community gardening hubs in the heartlands, NParks introduced the Enterprise Community Urban Farm (SECUF) Grant. This grant not only accelerates the setup of these hubs but also provides opportunities for micro-employment and volunteering for the less privileged and seniors. These initiatives have garnered overwhelming public response, as individuals recognise the positive health and welfare benefits associated with engaging in this form of walkability for a meaningful cause
	Communities Participation	A diverse group of volunteers, consisting of approximately retirees, homemakers, working professionals and students play a crucial role in transforming Singapore into a City in Nature. These dedicated individuals include members of Friends of the Parks, citizen scientists participating in the CIN initiative, and community gardeners involved in the CIB programme. Together, they assume a pivotal responsibility in creating a sustainable and vibrant natural environment for the benefit of all Volunteers serve as ambassadors and advocates, sharing their knowledge and experiences of living in a biodiverse and nature-rich city with communities. Their goal is to expand NParks' volunteer network to 70,000 by 2030 by engaging more schools and partners in reforestation efforts and citizen science projects. By increasing community participation, more people can actively practice walkability and access allotment gardens, while also promoting the cultivation of edible plants

Source NParks (2022)

The naturalistic landscapes at Bukit Gombak Park (a node in the Bukit Batok Nature Corridor, an ecological corridor between the Tengah Forest Corridor and Central Nature Park Network in western Singapore) bring the health and wellness benefits of greenery closer to residents while enabling walkability. Singapore's largest Therapeutic Garden in Jurong Lake Gardens of 3,100 square metres was the first to feature sections for children with conditions such as attention deficit hyperactivity disorder and mild autism as well as for seniors with dementia.

NParks has stepped up works to increase Singapore's biodiversity and to strengthen its ecological connectivity islandwide. Some 21 kilometres of Nature Ways hosting walking and cycling trails have been completed, primarily located in Kranji, Punggol Central, South Buona Vista and Upper Changi. NParks has naturalised 10 sites as a part of rewilding efforts by planting native trees to bolster ecological connectivity along Pasir Panjang Road and Singapore's expressways. The community can learn more about native species from volunteers at Invasive Species Management (ISM) workshops conducted at the Dairy Farm, Rail Corridor and Windsor Nature Parks.

Beyond physically planting trees, members of the community such as corporations, grassroots groups and schools are involved in propagating native tree saplings, designing planting plans, regenerating forests, education and outreach efforts, and initiating or facilitating events that contribute to the OneMillionTrees movement. Volunteers have been roped in for habitat enhancement works at eight parks including Changi Beach Park and Ang Mo Kio Town Garden West. Restoration of the coastal beach forest area between the sea and the coastal hill forest allows for more walkability space and is part of the Forest Restoration Action Plan for Labrador Nature Reserve, serving as an ecological buffer for the coastal hill forest to increase food sources for fauna. Pulau Satumu provides an opportunity for leisurely walks, as well as observing the dedicated efforts of volunteers and academic partners to study ways to manage various sea-based activities and minimise their impacts on corals with the aim of enhancing coral reproduction and ensuring the long-term sustainability of Singapore's marine biodiversity. The *Plant-A-Coral, Seed-A-Reef Programme* pilots the use of mineral accretion technology to enhance coral growth in Singapore's waters. NParks also harnesses nature-based solutions such as reforestation and redeveloping mangrove areas to boost Singapore's climate resilience. Over the years, citizen scientists have grown to 9,000, one of the larger groups of NParks' volunteer base of over 56,000. Community in Nature (CIN) Biodiversity Watches is a citizen science programme that garners community involvement in coordinated research efforts. The collection of data through CIN initiatives, with the assistance of citizen scientists, will provide NParks with valuable information about Singapore's biodiversity, enabling NParks to develop effective conservation strategies. CIN programmes are available to both school groups and the general public with programmes focusing on the marine environment, such as the Comprehensive Marine Biodiversity Survey and TeamSeaGrass. NParks also organises biodiversity watch such as the Dragonfly Watch, Heron Watch, Garden Bird Watch, and Butterfly Watch.

Nature is uplifted with NParks's effort to promote and intensify science and technology projects under NParks Science and Technology Masterplan. The works stretches from greenery management, biodiversity conservation to park and research. Flora of Singapore project is carried out by NParks in Bukit Timah Nature Reserve to identify genomic sequence of over 500 plant species. By building a more complete series of ecological, genomic and taxonomic studies, the study findings will inform flora conservation efforts on the different plant populations and Singapore's range of plant varieties. The project has resulted in the discovery of new records and rediscoveries of various algae, fungal and plant species. For instance, the *Hetaeria oblongifolia* orchid, which had not been seen for around 120 years was rediscovered, and the only known freshwater red algae species, *Visia cylindrocellularis,* was found, which is a new record for Singapore. NParks also sought to deepen its nature-driven research, offering a scientific pillar for Singapore's transformation into a *City in Nature*. The Cities of Tomorrow R&D Programmes in Research, Innovation and Enterprise was formed in 2021, leveraging on existing research and development efforts in biodiversity and greenery to inform nature-based solutions. Marine Climate Change Science (MCCS) programme is another project implemented by NParks that is primarily funded by the budget allocated under the National Research, Innovation and Enterprise 2025 Plan. The MCCS programme gathers scientific evidence and formulate interventions to mitigate the climate change impacts on marine ecosystems. NParks works with several MCCS members that map current and future research efforts in Blue Carbon Science, Climate Impact Science, Community- Driven Climate Resilience Planning, Eco-Engineering and Ecological Resilience. In a collaborative effort with the Department of Psychological Medicine at the Yong Loo Lin School of

Medicine at the National University of Singapore (NUS), a study has been conducted to design therapeutic gardens and landscapes in places like Bedok Reservoir Park and Pasir Ris Park. These spaces are curated based on evidence showcasing the positive impact of therapeutic environments on mental well-being and overall health. Access to these therapeutic gardens allows individuals to engage in self-care practices at their own pace through activities like walking, cycling and socialising. NParks has set a goal to establish 30 more therapeutic gardens throughout the city-state by 2030.

Clive Davies and his colleagues presented the notion that open spaces are a form of green infrastructure and should be assessed and designed based on three key dimensions—context (i.e., aspirations, needs, problems and wants), interaction (i.e., multiple functions connecting different elements) and quality (i.e., suitability and sufficiency).[15] These dimensions include various functions like conservation, creation, development, enhancement, initiatives and proposals. As shown in this chapter, Singapore's pursuit of becoming a City in Nature exemplifies this notion by placing great emphasis on urban greenery and making it a crucial aspect of Singapore's modern landscape.

While Singapore has cleared areas of forest for urban sprawl and reclaimed land from the sea to expand its coastline, Singapore is also home to a variety of animal species residing in nature reserves, marshes and wetlands. It is also worth noting that the city-state offers a plethora of exciting hiking trails winding through its hill reservoirs and forests. The trails play a crucial role in raising awareness about the importance of preserving natural ecosystems within urban areas. The Singaporean government recognises the significance of integrating urbanisation and natural ecosystems in a mutually beneficial way. This spatial reorganisation effort aims to create a harmonious coexistence between the urban core and the surrounding natural environment. It is a vital step towards realising Singapore's vision of becoming a City in Nature where Singapore's natural heritage is *enriched, revitalised* and *uplifted* while enabling walkability.

[15] Davies et al. (2006).

References

Benedict, M. A., & McMahon, E. T. (2001). *Green infrastructure: Smart conservation for die 21st century.* http://www.conservationfund.org/sites/default/files/GI_SC21C.pdf. Accessed on 21 December 2022.

Cameron, R. W., Blanuša, T., Taylor, J. E., Salisbury, A., Halstead, A. J., Henricot, B., & Thompson, K. (2012). The domestic garden—Its contribution to urban green infrastructure. *Urban Forestry & Urban Greening, 11*(2), 129–137.

Campbell, W. (1967, August 1). A new life planned for an old fortress. *The Straits Times*, p. 10. https://eresources.nlb.gov.sg/newspapers/Digitised/Article/straitstimes19670801-1.2.71. Accessed on 4 February 2022.

Connolly, C., & Muzaini, H. (2021). Urbanizing islands: A critical history of Singapore's offshore islands. *Environment and Planning E: Nature and Space.* https://doi.org/10.1177/25148486211051082

Corlett, R. T. (1995). Rain forest in the city: Bukit Timah Nature Reserve Singapore. *Gardens' Bulletin Singapore Supplement, 3*, 1–165.

Corlett, R. T. (1997). The vegetation in the nature reserves of Singapore. *Gardens' Bulletin Singapore, 49*(2), 147–159.

Davies, C., MacFarlane, R., McGloin, C., & Roe, M. (2006). *Green infrastructure planning guide.* Project: Final Report. https://www.researchgate.net/profile/Maggie-Roe/publication/265012095_GREEN_INFRASTRUCTURE_PLANNING_GUIDE_Authors/links/54f087a70cf24eb87940c8c1/GREEN-INFRASTRUCTURE-PLANNING-GUIDE-Authors.pdf. Accessed on 21 December 2022.

Ho, A. (2015). *Sentosa.* https://eresources.nlb.gov.sg/infopedia/articles/SIP_247 2005-01-20.html. Accessed on 4 February 2022.

Jerke, D. (2008). *Urban design and the bottom line: Optimizing the return on perception.* ULI-the Urban Land Institute.

Lai, S., Loke, L. H., Hilton, M. J., Bouma, T. J., & Todd, P. A. (2015). The effects of urbanisation on coastal habitats and the potential for ecological engineering: A Singapore case study. *Ocean & Coastal Management, 103*, 78–85.

Laurent, B., Doganova, L., Gasull, C., & Muniesa, F. (2021). The test bed island: Tech business experimentalism and exception in Singapore. *Science as Culture, 30*(3), 367–390.

Lee, M. T. G. (2000). *The making of Jurong Island—The right chemistry.* Jurong Town Corporation.

Marshall, S. (2004). *Streets and patterns.* Routledge.

Monteiro, R., Ferreira, J. C., & Antunes, P. (2020). Green infrastructure planning principles: An integrated literature review. *Land, 9*(12). https://doi.org/10.3390/land9120525

Ng, C. S. L., & Chou, L. M. (2018). Coral reef restoration in Singapore—Past, present and future. In *Sustainability matters: Environmental management in the anthropocene* (pp. 3–23).

NParks. (2022). *Sustainability report*. https://www.nparks.gov.sg/portals/annualreport/sustainability-report/index.htm. Accessed on 25 January 2023.

Public Works Department. (1975). *Trees are inspected regularly* (p. 52). Tee, S. P. (2014, June 23). Trees are inspected regularly. *Today*. http://tablet.todayonline.com/voices/trees-are-inspected-regularly

Tan, K. (2022). *A history of the Padang*. https://biblioasia.nlb.gov.sg/vol-18/issue-1/apr-to-jun-2022/history-padang. Accessed on 22 December 2022.

Tay, J. Y., Wong, S. K., Chou, L. M., & Todd, P. A. (2018). Land reclamation and the consequent loss of marine habitats around the Ayer Islands, Singapore. *Nature in Singapore, 11*, 1–5.

The Straits Times. (1969, January 23). A 'Paradise Island'. p. 4. https://eresources.nlb.gov.sg/newspapers/Digitised/Article/straitstimes19690123.2.20. Accessed on 4 February 2022.

The Straits Times. (1972, March 24). Sentosa pleasure island will have Pirates Cove. p. 32. https://eresources.nlb.gov.sg/newspapers/Digitised/Article/straitstimes19720324-1.2.110. Accessed on 4 February 2022.

Turnbull, C. M. (1972). *The Straits settlements 1826–67: Indian Presidency to crown colony*. Athlone Press.

CHAPTER 7

Managing Vehicle Population for Sustainable Development in Singapore

Sustainable development is a concept that has gained traction over the past few decades. It involves meeting the needs of the present generation without compromising the ability of future generations to meet their own needs. One of the biggest challenges in achieving sustainable development is managing the impact of human activities on the environment. One such activity that has a significant impact on the environment is the increasing use of private cars, which results in increased traffic congestion, air pollution and greenhouse gas emissions. The management of car population is crucial for achieving sustainable development.

The reduction of car usage also results in the reduction of the demand for fuel, thereby reducing the country's dependence on foreign oil. The money saved on fuel can be redirected towards the development of sustainable infrastructure, such as public transportation and cycling infrastructure. The emphasis on car population management therefore makes sense. Singapore is one of the countries that have been successful in managing car population. Singapore's car ownership rate is about 11% as compared to 50% in Europe and 80% in the United States.[1] Roads only occupy a small percentage of Singapore's land area (12%), in contrast

[1] Budget Direct Insurance (2023, February).

Melissa Liow Li Sa and Sam Choon-Yin, *Sustainable Urban Development in Singapore*, Sustainable Development Goals Series,
https://doi.org/10.1007/978-981-99-5451-3_7

to 14% for housing.[2] As part of the Land Transport Master Plan 2008, the vehicle growth rate was reduced from 3% in May 2009 to 0.25% in February 2015 and down to 0% in 2018. Goods vehicles and buses were granted an exception to the vehicle growth rate restrictions in Singapore where until 31 January 2025, a growth rate of 0.25% is allowed for these types of vehicles, as businesses require additional time to streamline their logistics operations and decrease the necessity for purchasing new commercial vehicles to sustain their operations.[3]

The objective of this chapter is to outline Singapore's initiatives to decrease the number of cars on the road and encourage the use of public transportation. The chapter also demonstrates how the design of MRTs can foster walkability and support sustainable development. To start, let us briefly examine Singapore's dedication to sustainability and its endeavours towards achieving this goal, particularly in regard to transportation management and its involvement with international organizations and ASEAN commitments.

Singapore's Sustainability Efforts and Its Involvement with International Organizations and ASEAN Commitments

The increase in the number of cars on the road can have negative impacts on the achievement of several SDGs related to sustainable cities and communities, climate action and clean energy. For example, SDG 11 aims to make cities and human settlements inclusive, safe, resilient and sustainable. However, the increase in the number of cars on the road can lead to congestion, air pollution and noise pollution, which can have negative impacts on the livability of cities and human settlements. SDG 7 aims to ensure access to affordable, reliable, sustainable and modern energy for all. However, the increase in the number of cars on the road can lead to an increase in the demand for fossil fuels, which are finite resources and contribute to greenhouse gas emissions. This can make it challenging to achieve the targets of SDG 7 related to clean energy.

Therefore, managing the car population is an essential component of achieving the SDGs related to sustainable cities and communities, climate

[2] Government of Singapore (2022).

[3] Ministry of Transport (n.d.).

action and clean energy. This can be achieved by promoting the use of public transportation, encouraging active transportation such as walking and cycling, and promoting the use of electric and hybrid vehicles. Additionally, policies such as congestion pricing and parking restrictions can help to manage.

International cooperation through organisations such as ASEAN and the UN can play a crucial role in achieving sustainable development goals related to the reduction of car population. At the global level, the United Nations Sustainable Development Goals (UNSDGs) provide a framework for countries to work together towards a sustainable future. The UN has set-up initiatives such as the Global Fuel Economy Initiative, which seeks to promote fuel efficiency standards for vehicles and encourage the adoption of cleaner and more sustainable modes of transportation, targeting the developing and transitioning countries. Through such initiatives, countries share knowledge and resources, develop common standards and policies and work towards a more sustainable future.

Indeed, Singapore's strategic developments and policies are mainly underpinned by UNSDGs under the 2030 Agenda for Sustainable Development (known as 2030 Agenda). Spearheaded by the Ministry of Sustainability and the Environment and the Ministry of Foreign Affairs, the Singapore government commenced its first Voluntary National Review of the SDGs at the UN High-Level Political Forum in July 2018, where its progress on sustainable development was shared with the delegates, predominantly in aspects of sustainable cities. In addition, Singapore Cooperation Enterprise (SCE), a joint initiative of the Ministry of Trade and Industry and the Ministry of Foreign Affairs, was established in 2006 to respond effectively to the constant foreign requests to tap into Singapore's experience in public governance, health, education, transportation, urban planning and others. The SCE charges fees to its clients. It typically begins by developing the Bilateral Technical Assistance Programme, which details the training initiatives, visits, consultancy workshops and implementation plans. SCE has worked with countries such as Brazil, China, Indonesia, Oman, Russia, Saudi Arabia and Vietnam.[4]

Singapore has been sharing its experience on sustainable development with developing nations through technical assistance under the Singapore Cooperation Programme (SCP), which was established in 1992. The

[4] For more details, see Calder (2017, pp. 144–145).

SCP offers a range of training programmes to share Singapore's development experience with other developing countries. The SCP Training Award provides an opportunity for mid- to senior-level government officials involved in public transport policy and planning to gain an overview of Singapore's urban transport system and develop skills to create and manage a sustainable, integrated and high-quality system. The programme aligns with SDG 9—Industry, Innovation and Infrastructure and covers topics such as master planning, regulatory framework development, fundraising for public transport, transport safety and cyber security. Participants learn to apply innovative technologies to increase transport efficiency and drive sustainable transport initiatives, such as Singapore's electric vehicle (EV) programme.[5]

The extraordinary scale and ambition of the 2030 Agenda entail rekindled pledge and greater partnership from all stakeholders. To monitor Singapore's development in attaining the SDGs, the Singapore Department of Statistics (DOS) unveiled an SDG webpage in September 2019. Singapore DOS at the regional front, is an active member of the ASEAN Working Group on Sustainable Development Goals Indicators (WGSDGI) which monitors and reviews the SDG indicators implemented in ASEAN. DOS was the chair of the WGSDGI and under their wing, the inaugural ASEAN SDG Indicators Baseline Report 2020 and the Online Database for ASEAN SDG Indicators in October 2020 were successfully launched, and in September 2022, the 2022 ASEAN SDG Snapshot Report was released.[6] ASEAN established the ASEAN Fuel Economy Roadmap, which aims to improve fuel economy standards for passenger cars and commercial vehicles across the region. Such initiatives demonstrate the willingness of ASEAN member states to work together to achieve sustainable development goals related to the management of car population.

Private Cars

Singapore is a small city-state that has consistently been ranked as one of the most liveable cities in the world. However, this has not always been the case. In the past, Singapore faced significant challenges in

[5] Singapore Cooperation Programme (2022, March 9).

[6] Government of Singapore (2022, December 8).

managing its car population due to rapid economic growth and limited land area. Facing a significant challenge with increasing traffic congestion and limited road space, the government implemented the Area Licensing Scheme (ALS) in 1975 where drivers were required to purchase a daily or monthly licence to enter the CBD during peak hours. The ALS was successful in reducing traffic congestion in the CBD by up to 40% and led to the implementation of other road pricing systems in the city-state.

In the early 1990s, the government introduced the Singapore Vehicle Quota System (VQS), which controls the number of new vehicles on the road. Under the VQS, a limited number of Certificates of Entitlement (COEs) are released through a bidding system each month, and buyers have to bid for a COE to own a vehicle. The price of the COE varies depending on the demand for and supply of COEs. This system has been effective in limiting the number of cars on Singapore's roads, as it makes car ownership more expensive and exclusive to those who can afford it. In addition to the COE system, the ERP system was introduced in September 1998, replacing the ALS for restricted zones to impose a fee on drivers for using certain roads during peak hours. This helps to reduce traffic congestion during rush hour and encourages drivers to use alternative modes of transportation.

Apart from the VQS, Singapore has implemented several other measures to discourage car ownership and encourage the use of public transport. One such measure is the Road Tax, which is based on engine capacity and emissions, and the Additional Registration Fee (ARF), which is a tax imposed on the open market value of a vehicle. The government has introduced initiatives such as the Vehicle Growth Rate (VGR) system, which outlines the number of new cars that can be registered each year. If the VGR is set at zero percent, it means that no new cars can be registered in Singapore unless an existing car is de-registered. In 2017, the government announced that it would cut the vehicle growth rate (VGR) to zero from 0.25% per annum for private cars and motorcycles, citing "land constraints" and a commitment to "continually improve our public transport system".

The Singapore Sustainability Blueprint was launched in November 2014 to turn Singapore into a car-lite society. At the launch of the Blueprint, Prime Minister Lee Hsien Loong told his audience, "We have to rely less on cars on the roads because we cannot keep on building roads—more roads for more cars. So, we will provide more options for Singaporeans that are better than cars. Buses—more of course. Expanding the MRT network—that is happening every day, but also other modes of transport, for example, bicycling".[7] With the right incentives to alter the behaviour, Kishore Mahbubani saw the possibility of reducing the demand for cars from 900,000 to 300,000.[8] The Singapore government's strategy to transform into a car-lite nation is a critical component of its long-term planning and infrastructure design. However, achieving this was a daunting task, as the public transport system prior to the 1960s was highly inefficient and disorganised. The bus routes were managed by private operators and frequently disrupted by strikes, leading to unreliable service quality. Additionally, severe traffic congestion posed a significant challenge, and the government had to prioritise the safety of the most vulnerable road users, i.e., pedestrians.

Overall, Singapore's past and present measures and policies to manage its car population have been successful in reducing traffic congestion and promoting sustainable development. While there may be challenges and limitations to these measures, the government's commitment to reducing the number of cars on its roads and investing in alternative modes of transportation is a positive step towards a more sustainable future.

Ride-Hailing Services

Singapore's ride-hailing services have a vital role in the country's transportation system, offering an alternative mode of transport to private cars and public transportation. From a sustainability perspective, ride-hailing services have the potential to contribute positively to reducing the environmental impact of transportation in Singapore.

[7] Speech delivered by Prime Minister Lee Hsien Loong at the Clean and Green Singapore 2015 Launch on 8 November 2015. https://www.pmo.gov.sg/Newsroom/transcript-prime-minister-lee-hsien-loongs-speech-clean-green-singapore-2015-launch-8. Accessed on 8 July 2022.

[8] Mahbubani (2017). In land scarce Singapore, reducing the demand for cars makes sense, considering that each car occupies two to six parking spaces at home, office and other destinations, which takes much more land comparatively speaking than walking and cycling.

Ride-hailing companies like Grab, Gojek, TADA and Ryde have a significant presence in the transportation market in Singapore.[9] It is encouraging to note that these companies are increasingly embracing clean energy in their operations. They are supporting ride-pooling and car sharing, as well as electrifying their vehicles. For example, Grab in Singapore has its own Transport Sustainability Goal, and committed to invest close to US$200 million to operate a mix of hybrid and electric vehicle fleet that runs on cleaner energy by 2030.[10]

The leading taxi operator in Singapore is ComfortDelGro, which operates over 13,000 taxis in the city. Other taxi companies include Transcab, SMRT and Premier Taxis, which offer a range of services, including standard taxis, limousines and airport transfer services. The taxi industry in Singapore is subject to regular checks and maintenance to ensure that vehicles are kept in good condition and are safe to operate. This helps to reduce the environmental impact of transportation by ensuring that vehicles are operating at their maximum efficiency and emitting fewer pollutants.

LTA Vehicular Emissions Scheme (VES) provides incentives to encourage taxi companies to switch to greener vehicles. The scheme offers rebates and imposes surcharges, depending on the vehicle's CO2 emission, plus emissions of 4 pollutants—hydrocarbons (HC), carbon monoxide (CO), nitrogen oxides (NOx) and particulate matter (PM). The pollutant with the highest emission value (i.e., the worst performing pollutant) determines the vehicle's band and its corresponding VES rebate or surcharge. From 1 January 2021 to 31 December 2023, newly registered electric taxis received a 45% rebate off the Additional Registration Fee (ARF), capped at S$20,000, and under the Enhanced VES framework, motorists who buy cleaner vehicles are entitled to increased concessions of up to S$25,000, while those purchasing more pollutive vehicles have to pay more. This incentivizes taxi companies to opt for hybrid and electric cars that emit fewer pollutants, leading to a significant reduction in carbon emissions from the transportation sector.

[9] Ride-hailing services contribute to about 4% of Singapore's land transport emissions. Chan, A. (2021, September 6). *Commentary: How ride-hailing will change to help countries meet climate goals.* https://www.channelnewsasia.com/commentary/grab-transport-sustainability-uber-didi-app-ride-driver-ev-net-zero-2154676#:~:text=While%20ride%2Dhailing%20accounts%20for,US%24230%20billion%20by%202026.

[10] Kok (2021, July 7).

In addition to promoting the use of greener vehicles, ride-hailing services in Singapore are subject to regulations that require them to adhere to fuel efficiency standards. The LTA sets fuel efficiency targets for taxi companies, with the aim of reducing carbon emissions from the transportation sector. This has resulted in the introduction of more fuel-efficient vehicles and the adoption of practices such as eco-driving to reduce fuel consumption and emissions.

Public Bus Services

Public transportation such as public bus services helps to reduce greenhouse gas emissions and air pollution by reducing the number of cars on the road. When more people take public transport, traffic congestion is reduced, leading to time and fuel savings. Moreover, demand for parking space is expected to decrease, which can lead to reducing the amount of land needed for parking lots and garages and promote the development of more sustainable urban areas. In Singapore, the public transportation system is one of the best in the world, with an extensive network of bus services and Mass Rapid Transit (MRT) trains.

The history of the bus system in Singapore dates to the 1920s when privately-owned buses were the only form of public transportation. In 1933, the Singapore Traction Company (STC) introduced its first bus service, which was a great success. The STC eventually became the Singapore Bus Service (SBS) in 1973, which was later renamed as the SBS Transit in 2001. The government also established a second bus company, the Trans-Island Bus Service (TIBS), in 1982, which was merged with the SBS Transit in 2001.

In the early years, the bus system was plagued by problems such as poor service quality, inadequate coverage and overcrowding. However, the government took a series of measures to improve the system. Today, the bus system in Singapore is known for its high service standards and reliability, with an average daily ridership of four million trips.[11] The buses operate on a scheduled basis, with high-frequency services during peak hours and lower frequency services during off-peak hours. The government also invested heavily in the system's infrastructure, such as building

[11] LTA (2021, p. 30).

new air-conditioned bus interchanges, improving bus stops and expanding the road network.[12]

The bus system in Singapore has been designed to reduce the environmental impact of transportation and promote sustainable living. Firstly, the bus system in Singapore helps to reduce air pollution. This is because buses run on compressed natural gas (CNG) and low sulphur diesel, which are cleaner and emit fewer harmful pollutants into the air compared to conventional diesel buses. This is significant in Singapore, where the air quality is a major concern due to the country's urban landscape and high population density. By reducing the number of conventional diesel buses on the road, the bus system helps to mitigate air pollution and promote a healthier environment for the people of Singapore.

Secondly, the bus system in Singapore helps to reduce carbon emissions through the implementation of bus lanes and priority measures for buses, which reduce the amount of time buses spend on the road and their fuel consumption. The bus lanes and priority measures for buses also help to speed up bus journeys, making public transportation a more attractive option for commuters. This encourages more people to use public transportation and reduces the number of private vehicles on the road, which in turn reduces traffic congestion and helps to mitigate the negative environmental impacts of private vehicle usage.

Thirdly, the introduction of electric buses in recent years has been a significant step towards reducing the environmental impact of the bus system in Singapore. LTA has implemented a trial to switch to cleaner public businesses to reduce their carbon footprint. The goal is to have a 100% cleaner bus fleet by 2020. As reported in LTA's Annual Report "LTA has deployed 50 diesel hybrid buses (DHB) and will progressively deploy 60 electric buses. The experiences gained from the deployment of these cleaner energy buses will prepare Singapore's transportation system to tackle the challenges that come with large-scale EV deployment in the future. An electric bus is expected to reduce carbon emissions by 60%, relative to a diesel bus (after accounting for grid emissions)".[13] This shows the commitment of the government of Singapore towards promoting sustainable transportation options.

[12] Bus interchanges are strategically located and serve as the main transfer points for buses. For excellent reviews of Singapore's bus system, see Spencer (1988) and Leong et al. (2016).

[13] LTA (2021, p. 19).

Fourthly, by covering almost every part of the island, the bus system in Singapore provides accessibility to various neighbourhoods and facilitates the movement of people within the city. This helps to promote a compact and well-connected city, which reduces the need for long-distance commuting and promotes a more sustainable and livable environment for the people of Singapore. With the bus system integrated with other modes of public transportation such as the MRT system, it provides a comprehensive and efficient transportation network that reduces the need for private vehicles.

In sum, the bus system in Singapore plays a significant role in reducing the environmental impact of transportation. It helps to reduce air pollution, carbon emissions and traffic congestion while promoting sustainable urban development. The bus system in Singapore serves as a model for other cities to follow, as it demonstrates the feasibility of reducing the environmental impact of transportation through the promotion of sustainable practices.

Green Transit-Development (TOD) and MRT Stations

The Singapore government has invested heavily in the development of trains, with the aim of providing Singaporeans with a reliable, efficient and affordable alternative to car ownership. The MRT and LRT systems are the backbone of Singapore's public transportation system, and they provide an affordable and convenient way to travel around the city-state.

The MRT system has been in operation since 1987 and is managed by LTA. The MRT system initially had five lines, and over the years, new lines have been added to meet the growing demand for public transportation. The LRT system is a newer addition to Singapore's public transport network. The first LRT system, the Bukit Panjang LRT, was opened in 1999, followed by the Sengkang-Punggol LRT in 2003. The LRT system is designed to serve areas that are not well served by the MRT system and to provide last-mile connectivity to residents in these areas. The MRT system, in particular, has been a game-changer, providing a fast, reliable and affordable alternative to driving. According to the LTA, in between January and December 2022, the MRT system had an average daily ridership of between 2.16 and 3.19 million, while the LRT system had an average daily ridership of 160 to 190 thousand commuters.[14] These numbers indicate the popularity of the MRT and LRT systems and their effectiveness in reducing the usage of cars on the roads.

The MRT construction process typically begins with feasibility studies, which assess the potential demand, technical feasibility and environmental impact of the proposed MRT line. Once the feasibility studies are completed, the government will secure funding and appoint a construction contractor to design and build the MRT infrastructure. The construction process of MRT involves various stages, including excavation and tunnelling, station construction, track laying and system installation. Construction sites are carefully managed to minimise disruption to the surrounding environment and community, with noise and dust levels closely monitored and mitigated.

The construction process considers Green Transit-Development (TOD) concept which promotes the creation of vibrant, walkable neighbourhoods with access to a variety of amenities and services, such as parks, shops and schools. TOD developments typically include a mix of residential, commercial and office spaces, with the aim of reducing commuting times and encouraging sustainable modes of transport, such as walking, cycling and public transport. Green TOD underlines the managed development of society, transportation and the environment, to construct a sustainable and liveable transit-oriented community. In other words, new environment and social dimensions are incorporated in the conventional TOD by combining green urbanism and ideas related with an ecological community. Green TOD is an advancement of the TOD theory that emerged from the themes of green urbanism and sustainable development.

In their work, Robert Cervero and Cathleen Sulivan emphasise the following aspects as essential in Green TOD: (1) mixed land use, (2) higher densities, (3) decrease surface parking and replacing impervious surfaces with pocket parks and community gardens as well as (4) solar energy production at stations.[15] Aspects (1), (2) and (3) are directly linked to the built environment of the transit station areas, and aspect (4) is more applicable from the viewpoint of ecological technology and green architecture. Mixed land use brings destinations closer but also

[14] Government of Singapore (2023, March 8).

[15] Cervero and Sullivan (2010).

forms a vibrant community life. Surface parking restored by spaces with pocket parks and community gardens that would yield social and environmental benefits. Combining green urbanism and TOD can reduce carbon emission and energy usage up to 30%. Correspondingly, Green TOD put emphasis on the synchronised development of society, transportation and the environment, to foster a sustainable and liveable transit-oriented community.

The concept is relevant to Singapore. Over the years, Singapore's city planning had been developed around its MRT stations, to a large extent combining green urbanism with TOD to generate broader environmental, ecological and social benefits, while promoting walkability and cycling. Niu Shaofei, Hu Ang, Shen Zhongwei, Huang Ying and Mou Yanchuan's study present a systematic and quantitative investigation of the built environment of the Green TOD by using 13 variables to measure the built environment of 23 sample MRT station areas in Singapore.[16] To analyse the built environment factors of Green TOD, a k-means cluster analysis was conducted based on five factors, namely the neighbourhood spatial form factor (Factor 1), facility completeness factor (Factor 2), open space factor (Factor 3), high-density mixed land-use factor (Factor 4) and walking amenity factor (Factor 5). The purpose of this analysis was to identify similar subgroups among the 23 MRT stations in terms of their built environment properties. The analysis resulted in the division of the 23 MRT stations into five distinct clusters (Table 7.1).

The study's findings revealed several issues related to the quality of public space and connectivity, as well as the inadequacy of walking conditions in certain areas in Cluster 1 (Bedok, Bishan, Bukit Panjang Hougang, Pasir Ris, Sengkang, Serangoon) compared to other clusters. Most of these areas are located in developing towns, such as Pasir Ris and Bedok stations, which suffer from a lack of public space and connectivity. For instance, Pasir Ris station is surrounded by a large area of reserved land, while Bedok station has the largest business land among the 23 stations but the shaded walkways and public space such as office park are ineffectually configured. Additionally, there were inadequate pedestrian

[16] Niu et al. (2021). The 13 variables refer to (1) building density (2) average plot ratio (3) entropy index of land-use mix (4) proportion of commercial land to residential land (5) average block length (6) street intersection density (7) walkway length (8) shaded walkway length (9) number of bicycle racks (10) area of parks and open spaces (11) open space connectivity (12) number of public space facilities (13) number of bus stops.

Table 7.1 Five dimensions of Green TOD in 23 MRT stations

Five dimensions of Green TOD	*23 MRT stations across 5 clusters*
Factor 1—Neighbourhood spatial form factor **Factor 2**—Facility completeness factor **Factor 3**—Open space factor **Factor 4**—High Density mixed land-use factor **Factor 5**—Walking amenity factor	**Cluster 1**: Bedok, Bishan, Bukit Panjang Hougang, Pasir Ris, Sengkang, Serangoon **Cluster 2**: Jurong East **Cluster 3**: Paya Lebar, Punggol, Tampines, Woodlands, Yishun **Cluster 4**: Boon Lay, Bukit Batok, Choa Chu Kang, Kallang, Redhill **Cluster 5**: Ang Mo Kio, Clementi, Queenstown, Sembawang, Toa Payoh

Source Niu et al. (2021)

facilities like shaded corridors. The study found that all five factors related to TOD efforts require improvement. As a result, the study recommends that more parks and open spaces be added and that further improvements be made to pedestrian quality in neighbourhoods. These recommendations aim to enhance the quality of public space and connectivity, as well as to improve the walking conditions in these areas. By implementing these recommendations, it is hoped that these neighbourhoods will become more attractive and accessible to residents and visitors alike.

Cluster 2 (Jurong East) possesses TODs featuring high-density mixed-use and walkability. The study findings show that the Jurong East station area is characterised by good walkability and connectivity amidst a high-density mixed-use development in the core site. It exemplifies an urban complex group spanning neighbourhoods, with an average block length of about 240 m. The station is an interchange for the East-West and North-South lines of Singapore and is a regional centre in the western region. The area has small residential functions since the land surrounding the station site is slated for business and commercial functions. The elevated pedestrian network of J-Walk has greatly enabled accessibility among land parcels, forming a close linkage between Jurong East station, residential, commercial (J-Cube, IMM, Westgate) and medical care institutions (Ng Teng Fong General Hospital). Overall, the study finds that Cluster 2, which includes the Jurong East station area, possesses TODs featuring high-density mixed-use and walkability. The highest value is ranked for Factor 4 (i.e., the high-density mixed land-use factor) and Factor 5 (i.e., the walking amenity factor). The lowest value is ranked for

Factor 1 (i.e., the neighbourhood spatial form factor). The study recommends an increase in the number of bus stops and the length of park connectors in the Jurong East station area to improve the TODs further.

Cluster 3 (Paya Lebar, Punggol, Tampines, Woodlands and Yishun) was evaluated for Green TOD performance. Findings show that Factor 2, which refers to facility completeness, was rated low compared to Clusters 1, 2, 4 and 5. However, Factor 2, the neighbourhood spatial form factor, was rated highest, and the comprehensive factor scores of these five MRT stations in this cluster were rated top at all sites, suggesting better Green TOD performance. Cluster 3 features small blocks with inadequate facilities, such as Punggol in the northeast of the Singapore island, which was designed based on ecological and green concepts. The Punggol MRT station connects the light rail transit (LRT) lines and represents a checkerboard pattern of the road grid. However, there are fewer public service facilities in the Punggol MRT station compared with other MRT stations.

Another example is the Tampines station, which is elevated to keep a distance from the traffic routes. Pedestrian core spaces are found on both sides of the station, with an average block length of 160 metres surrounding the Tampines station. Shopping centres, office buildings and a bus interchange are integrated into the station exits, and there are retail and service functions in the spaces in the station viaducts. There are still many undeveloped land sites, and the design of accessibility of public service facilities is distributed in the neighbourhood centres with the adjoining LRT stations rather than arranged in the town centre. The study recommendation increasing the range of public service facilities like community centres, educational institutions and medical care, to enable more walkability and cycling to further improves the Cluster's TOD performance.

Cluster 4 (Boon Lay, Bukit Batok, Choa Chu Kang, Kallang, Redhill) features TODs with low development rate and lots of open space that have fairly complete nearby parks and longer park connectors. The study found that the Cluster had a low development rate and a significant amount of open space. In this cluster, Factor 3 (the open space factor) was rated highest while Factor 4 (the high-density mixed land-use factor) was ranked lowest compared to other clusters. Bukit Batok station, located in the town centre of Bukit Batok, had two neighbourhood parks within 500 metres of the station, which accounted for 19.3% of the green space. Overall, the cluster had a lush green space characterised by nearby parks and longer park connectors. However, the ratio of commercial to residential functions was relatively low compared to other station areas. To optimise the environmental benefits of Green TOD, the study recommends adopting an optimization strategy to increase the range of functions in the core area and develop a livelier town centre to further promote walkability and cycling.

Cluster 5 (Ang Mo Kio, Clementi, Queenstown, Sembawang, Toa Payoh) consists of MRT stations located in completely developed towns with rich and comprehensive public service facilities and bus stops, resulting in the highest mean value of Factor 2. One example is Clementi in the southwest of Singapore, which has a variety of public service facilities, such as schools, hospitals, libraries and shopping centres, in the station area. The Clementi Mall, an urban regeneration project seamlessly connects to the Clementi MRT station and adjoining neighbourhoods. The mall complex includes three residential towers, service and shopping areas and a bus interchange station. Additionally, Clementi Plaza, located south of the Clementi MRT station, provides a unique sense of community identity through its retail market and hawker centre. Overall, the MRT stations in Cluster 5 serve as both a transportation hub and a public activity space for the surrounding communities. To maximise the walkability of station areas in Cluster 5, the study recommends creating a more pleasant environment for pedestrians by increasing pedestrian and cycling facilities, based on the morphological assessment and factor scores.

It is evident that Green TOD plays a crucial role in promoting safe, inclusive, resilient and sustainable human settlements in Singapore. However, the five clusters have varying levels of development. The Green TOD built environment emphasises multifactor connection and integration. Among the five clusters, Cluster 1 has the worst development status in its TOD efforts, while Cluster 2 has the best high-density mixed-use development and walkability infrastructure. Cluster 3 needs to improve its facility completeness, whereas Cluster 5 has excelled in this aspect. For Cluster 4, which features TODs with low development rates, walking and cycling are facilitated due to its parks and longer park connectors.

The five factors of the TOD stations in the five clusters are closely related to regional development planning, town development phase and the spatial structure of the station area. The execution strategies of the TOD concept reflect the performance of the built environment. Cluster 1 requires more public spaces and connectivity, including more parks and open spaces. In Cluster 2 and Cluster 4, increasing the number of bus stops and the length of park connectors would encourage more walking and cycling. Cluster 3 and Cluster 4 need a wider range of public service facilities and a more vibrant town centre respectively. Cluster 5 has well-established towns with comprehensive public service facilities. Improving pedestrian and cycling facilities within a pleasant environment are important considerations to advance Singapore's Green TOD efforts.

Overall, it can be argued that the MRT stations located in HDB towns provide seamless and affordable transport options, including walkability and cycling, from residential areas to the stations. The proximity of community centres, educational institutions and medical care facilities to the MRT stations means that residents have access to important public services. These measures promote sustainable urban growth and are in line with SDG11—Sustainable Cities and Communities, as they enable residents to lead a high-quality lifestyle with convenient access to public transportation. Transit villages, which have emerged in the HDB towns around Singapore's MRT station centres, serve as hubs for commerce, food services, education and healthcare—all key components of sustainable urban development.

As we have shown in this chapter, managing car population has been a top priority for the government to ensure sustainable development and an efficient transportation system. Singapore's car population has been managed through a combination of measures, including regulatory policies, financial incentives and infrastructure development. The government has also invested heavily in developing a reliable and efficient public transportation system. The MRT and LRT systems are the backbone of Singapore's public transportation system, and they provide an affordable and convenient way to travel around the city-state. The government has implemented a comprehensive bus network that covers almost every part of the city-state. The public transportation system in Singapore is efficient, affordable and reliable, making it a popular choice for commuters.

The 2020 Census of Population report indicates that the initiatives to manage the number of cars in Singapore have been effective. According to the report, slightly over one-fifth of employed residents commuted to work by car only, down from about a quarter in 2010.[17] Additionally, the proportion of those who travelled by taxi or private hire car only increased from 1.3% to 3.0% over the same period. The use of public bus and MRT/LRT remained popular for employed residents commuting to

[17] Singapore Department of Statistics (2021).

work. In 2020, 57.7% of employed residents used public bus, MRT/LRT or a combination of both, which is an increase from 54.6% in 2010. This increase was due to a rise in the proportion of those who commuted to work by MRT/LRT with a transfer to or from public bus, in response to the expansion of the MRT/LRT network.

References

Budget Direct Insurance. (2023, February). *Car Ownership Singapore 2023.* https://www.budgetdirect.com.sg/car-insurance/research/car-ownership-singapore. Accessed on 25 March 2023.

Calder, K. (2017). *Singapore: Smart city, smart state.* Brookings Institution Press.

Cervero, R., & Sullivan, C. (2010). *Toward green TODs.* UC Berkeley Center for Future Urban Transport.

Government of Singapore. (2022, December 8). *Sustainable development goals.* https://www.singstat.gov.sg/find-data/sdg. Accessed on 26 March 2023.

Government of Singapore. (2023, March 8). *Statistics public transport ridership.* https://www.lta.gov.sg/content/ltagov/en/who_we_are/statistics_and_publications/statistics.html . Accessed on 25 March 2023.

Kok, Y. F. (2021, July 7). Grab to trial new green ride-hailing service. *The Straits Times.*

Leong, W., Goh, K., Hess, S., & Murphy, P. (2016). Improving bus service reliability: The Singapore experience. *Research in Transportation Economics, 59*, 40–49.

LTA. (2021). *Greener commutes, sustainable lifestyles annual report.* LTA.

Mahbubani, K. (2017). Singapore: The smartest city on our planet. In C. K. Heng (Ed.), *50 years of urban planning in Singapore.* World Scientific Publishing Corporation.

Ministry of Transport. (n.d.). *Sustainable Land transport policies.* https://www.mot.gov.sg/what-we-do/green-transport/sustainable-land-transport. Accessed on 25 March 2023.

Niu, S., Hu, A., Shen, Z., Huang, Y., & Mou, Y. (2021). Measuring the built environment of green transit-oriented development: A factor cluster analysis of rail station areas in Singapore. *Frontiers of Architecture Research, 10*, 652–668.

Singapore Cooperation Programme. (2022, March 9). *Integrated urban transport planning and management.* https://scp.gov.sg/startpublic/#!/courses/Search/transport/8685/Integrated%20Urban%20Transport%20Planning%20and%20Management/5/CLS_3. Accessed on 26 March 2023.

Singapore Department of Statistics. (2021). *Population trends 2021.* https://www.singstat.gov.sg/-/media/files/publications/population/population2021.ashx. Accessed on 8 May 2022.

Spencer, A. H. (1988). Modernisation and incorporation: The development of Singapore's bus services 1945–1974. *Environment and Planning A: Economy and Space, 20*(8), 1027–1046. https://doi.org/10.1068/a201027

CHAPTER 8

Urban Design for the Economy

When Singapore gained independence in 1965, the Singapore economic model was one of courting multinational corporations to set-up low-cost manufacturing in Singapore to create jobs and export to the world. The model was unconventional as the key theories of development at that time called for import substitution and protection of local markets. Yet, the model has proven to be successful for Singapore, and adopted by many others later on, including China's Deng Xiaoping following China's reform and opening up initiatives in the late 1970s.

In the early years of independence, jobs were widely available in the manufacturing sector. The workers form the working class, including people in production operations, moving materials, repair, maintenance and repair work. Later on, jobs in the services sector such as hospitality, transportation, sales and heath care expanded, including people in low wage sectors like health care, retail, clerical and administrative work, personal care and food services.

A third category of workers, the creative class, has gained prominence. The creative class, comprising scientists, engineers, university professors, poets, novelists, entertainers, designers, actors, artists, architects, analysts, web designers, software programmers, composers, writers, editors and other professionals in knowledge-intensive industries such as financial services, legal, health and high-tech sectors, is the driving force

Melissa Liow Li Sa and Sam Choon-Yin, *Sustainable Urban Development in Singapore*, Sustainable Development Goals Series,
https://doi.org/10.1007/978-981-99-5451-3_8

of economic growth. Typically drawing on complex branches of knowledge to solve specific problems, the creative class is highly valued and the associated traits like innovativeness, problem solving and analytical are important soft skills to acquire in schools and institutions of higher learning. According to Richard Florida, the creative class accounts for about one-third of the United States workforce or 40 million Americans in the early twenty-first century up from roughly three million in 1900, and will continue to climb upward in number, a reflection of the general process of economic and social change.[1]

An increasingly influential view, associated with scholars like Jane Jacobs and Richard Florida, is that innovative persons are more likely to base themselves in cities which are able to offer a creativity climate. The creative class, wrote Richard Florida, is oriented to cities which offer a stimulating and open environment where a person can freely express himself and tap into social networks to spur innovation and ideas, citing examples of leading creative cities like Washington D.C., Boston, Seattle and Austin. Jane Jacobs's *The Death and Life of Great American Cities* documented how this worked in Hudson Street where she lived. The presence of various types of buildings—shops, small factories, apartments, entertainment outlets and restaurants—within close proximity facilitate conversation and collision of ideas.[2] Wide sidewalks in particular provide the venue for different kinds of people to converse and interact and gather for relaxation and generation of ideas. The creativity connection, argued Jacobs, require an appropriate physical environment. What this also means is that urban planning is related to economic planning and development. Urban planning supports the movement of people, affects the quality of life and influences the decision to migrate especially among the creative and innovative class.

As a city-state which depends highly on human capital to further its prosperity, it makes sense for Singapore to strive to offer liveable city which people want to retain its competitiveness. Singapore's message to the creative class is that Singapore is the best place to work, live and play.

[1] Florida (2002a).

[2] Jacobs (1972).

Economy First

Singapore is often regarded as an economic success, having transformed from a low-income economy in the 1960s, in what is often described as an economic miracle. Promoting economic growth and creating jobs for the people are the key missions of the elected government. Lee Hsien Loong, as the then Deputy Prime Minister of Singapore, once asserted "If we ask the people to choose between more freedom, democracy and more economic growth, we have no doubt that they would choose economic growth.' But we will never ask them anyway."[3] The remark illustrates Singapore's pre-occupation with economic development, making everything subservient to economic prosperity. That bilateral relations are driven by material interests, at least from the Singapore side, is a common perception. The former Indonesian Ambassador to Singapore alleged that "Singapore tends to be self-centred, lacks empathy and is highly materialistic, is not a warm-hearted personality and therefore views Indonesia only in terms of 'profit or no profit'".[4]

The 'economy first' mentality prevails in Singapore for a reason. When the PAP took over the governance of Singapore in 1959, the economy was experiencing a high unemployment rate (of 5%), rampant corruption, a high crime rate, a large budget deficit (of US$4.7 million), rapid population growth and a public housing problem (50% of the population were living as squatters).[5] Faced with the prospect of the infiltration of communist ideology in the early years, modern social policies and rapid economic development were perceived as the main ingredients to banish the danger of Communism by creating employment opportunities. In 1961, the EDB was established to plan, coordinate and direct Singapore's industrialisation programme, which served as the main force for the nations' economic growth. Jurong Island Estate, a 5,000-hectare area in the western region of Singapore, was developed to house industrial factories to provide large-scale employment.

The key investment arms of the Singaporean government (Temasek Holdings and Government of Singapore Investment Corporation, GIC) have been conceptualised as strategic institutions to safeguard Singapore's

[3] *The Straits Times*, 9 July 1991 (quoted in Birch, 1994, p. 2).

[4] Abin (1991, p. 99).

[5] Quah (2008).

assets against the challenges faced by the small country.[6] This effectively legitimises economic wealth generation as the primary goal of the institutions. The Temasek Charter, published in 2002, conveys the message that Temasek's mission is to achieve sustainable *economic* returns for shareholders.

Tree planting exercise, an initiative started in November 1971, and nature conservation efforts could be described as economically motivated. As stated by the first Prime Minister, "in wooing investors, even trees matter".[7] A clean and green Singapore helps to enhance economic performance through the arrival of tourists, 'green' companies and foreign investors attracted to a well-organised place, although the initiative also helps to conserve the environment by removing carbon dioxide from the atmosphere. On nature conservation, the real motivation for the government's adoption of the 87-hectare Sungai Buloh as a nature park lies with the necessity to derive political legitimacy through continual economic development. "Being environmentally sound and cleaning up after rapid industrialization", argued Daniel Goh, "prevents the erosion of its (the government's) legitimacy base, as adverse environmental degradation would be seen as stemming from the same state-engineered economic development".[8] Indeed, in an economically defined national identity nation state, nature conservation reflects an opportunity cost at a time when Singapore seeks for wealth in an increasingly globalised and competitive world.

That a robust economy is needed to ensure Singapore's survival as a nation is a well-accepted premise. This is illustrated in the Singapore government's national report for the 1992 United Nations Conference, which stated that "the principles guiding Singapore's environmental policies can be summarised in this order of priority: (a) satisfy the people's economic needs first; (b) control population growth; (c) prevent pollution at source; (d) conduct Environmental Impact Assessment (EIA) for all development projects...".[9] The nature of the environmental problem—notably burning of fossil fuels for commercial and household usage and

[6] Clarke and Monk (2010).

[7] Senior Minister Lee Kuan Yew on the 35th Anniversary of the Economic Development Board on 1 August 1996. Quoted in Lye (2008, July, p. 73).

[8] Goh (2001, p. 12).

[9] United Nations Conference on Environment and Development (1992, p. 9).

reusing and recycling of products—implies that restraint has to involve changes in behaviour of the population, and not merely the public sector and the state. Thus, a responsible government may be seen as an important part of protecting the environment by getting the message across and receiving the support from the public.

Urban planning and economic planning are therefore related and connected. Economic planning facilitates the activities and movement of people to support the goals as identified in the economic plans. The case of attracting and retaining talent with particular consideration for urban design is discussed in the following section.

Talent Development

But first, what is talent? The term 'talent' can refer to the special ability, knowledge or skills that someone has to do something well and, in the process, contribute to the organisational success in the marketplace. Examples of talent include artistic talent like acting, painting, singing and graphic design; sporting talent like balancing, running and swimming; and academic talent like analytical and creating thinking and problem solving. We can do something to be more skilful and knowledgeable, for example, through formal training and education. Organisations around the world have in place talent development programmes to tackle and deal with succession planning issues.

To prepare for the future of work, the education sector has started to strive to improve student problem solving ability and promote the development of curiosity and ingenuity. Students are encouraged to come up with ideas, creative ideas. There is much determination to redirect their efforts from academic subjects towards skills-based curriculum, focusing on skills such as team building, public speaking and problem solving.

Possessing the traits is expected in the workplace. At work, individuals are expected to work together in a complex social system. They have to respond quickly, solving complex and interdisciplinary, open-ended problems, making innovative use of information, knowledge and opportunities. Organisations around the world understand that the bottom-line success relies on the ability to innovate to deliver the best product, process and service, and that before any innovation can take place, the ideas must be generated by individuals and teams, not machines. We need the raw mix of interrelated knowledge. We need schools and educational leaders to recognise the need for less predictable human thinking, giving the emotional, imaginative and cognitive lift that cannot be ignored in the digital age.

In Singapore, universities traditionally separate the study of humanities and sciences. But they are experimenting with more flexible ways of learning, recognising that interdisciplinary study is needed to find solutions to multifaceted problems. Nanyang Technological University and Singapore Management University offer a common core curriculum for their undergraduate students. The modules would take up to a third of the academic workload, integrating skills and knowledge in communication, digital literacy, enterprise and innovation. Singapore University of Technology and Design (SUTD), despite its focus on technology, requires students to take subjects in humanities and social sciences. Singapore's oldest university, National University of Singapore (NUS), would bring together the Faculty Of Arts And Social Sciences and Faculty Of Science to form a College Of Humanities And Sciences. Tan Eng Chye, President of NUS, noted that young adults today need to grapple with "wicked problems", problems that are ill-defined and multidisciplinary. As such, NUS "will ease the excessive requirements for subject majors so that we can add breath to the curriculum overall. If we want our students to be able to connect the dots in new and unusual ways, it helps to broaden their horizons and expose them to more dots".[10] Singapore Management University (SMU) focuses on developing the entrepreneur mindset; students who will think out of the box, innovate and be willing to take risks and try new things. Lily Kong, President of SMU, believed that one can be an entrepreneur whenever he is, "running our own business, working in a big business or a small company, and you can also be an entrepreneur in government. It's really about the mindset". SMU has set-up incubation space for start-ups to provide a platform to foster innovation and entrepreneurship at SMU and offer students more global experience to seek opportunities and immerse in projects with centres around the world. These are in addition to offering courses in entrepreneurship.[11]

To retrain workers, to strengthen capabilities of Singaporeans, the government initiated the SkillsFuture movement, a nationwide strategy to develop skills that are relevant to the future with detailed description

[10] Tan (2020, September 10).

[11] Singapore Management University (2019, March 25).

of skills at various positions, to enable Singaporeans to apply knowledge and skills in real-world situations that are constantly changing through practice and collaboration. SkillsFuture empowers Singaporeans to take charge of own skills, to pursue interests, aspirations and passions and reap a sense of joy and fulfilment. Singaporeans are given S$500 credit to pay for a wide range of courses from languages to cooking.

There is clear evidence of efforts by the government to train and retrain students and the workforce to prepare for the future of work. The key challenge is to retain the workforce, both locals and foreigners to continue contributing to the Singapore economy. One of the most significant features of this group of creative and innovative class is the shift aways from traditional and conservative places towards cities and regions that offer not just opportunities, amenities and financial rewards, but stimulating environment where they can mix work with play, express freely and interact with new and different kinds of people to facilitate the generation of knowledge and ideas.

Attracting Foreign Talent

Singapore has traditionally been the preferred destination for labourers from China and India. With a falling fertility rate in an ageing society, Singapore experiences shortage labour in many sectors of the economy and continues to welcome foreign workers, especially those who are highly skilled. This is evidenced in the 2000s when Singapore shifted its focus towards knowledge and innovative intensive industries in biomedical science, digital technology and engineering.

The Agency for Science, Technology and Research (A*STAR), previously known as National Science and Technology Board (NSTB), was established in January 2002 to advance Singapore's research and development capabilities by attracting highly skilled people and the nation into an innovation hub. A*STAR adopts a two-pronged strategy. First is to identify committed young scientists and create a way to fund their studies up to the doctoral level. The government-funded National Science Scholarship was setup in 2001 to achieve this. Second, A*STAR strives to recruit internationally acclaimed scientists to Singapore to lead cutting-edge research projects. Their presence as the argument goes would raise the profile of Singapore as a serious research and development hub, which in turn would attract global talent to Singapore as a place of employment.

Non-residents in Singapore typically hold one of the following passes.

- The P and Q Pass, or employment pass (EP Pass) for professionals, managers and administrators with a minimum basic salary of S$3,300 and a degree or other professional qualification
- The S Pass for mid-level professionals and technicians with a minimum basic salary of S$2,200
- The R Pass or work permit for unskilled or semi-skilled workers with a basic salary of less than S$2,200

In 1970, foreign workers accounted for 3.2% of the total workforce. The figure has risen to 7.4% and 37.8% by 1980 and 2013, respectively.[12] About 40% of the total population in Singapore was foreign born. The influx of migrants to the city-state has led to crowded public places such as eating places and public transport, competition for jobs and the impact on price levels, including housing, appear to be the concerns raised by the majority. The release of the Population White Paper in January 2013, which projected a total population of between 6.5 and 6.9 million by 2020—of which citizens would constitute only 55% (3.6 to 3.8 million)—unleashed a torrent of public criticism. Four protests were organised between 2013 and 2015 at Hong Lim Park. The largest anti-immigrant protest of 4,000 people was held on 16 February 2013. Dissatisfaction with the government's open-door policy had been identified by the ruling party's supporters as one of the sources of the party's less-than-satisfactory performance at the 2011 General Election where it won 60.1% of the popular votes from 66.6% in 2006. The oppositions won six parliament seats, the highest since independence.[13] The influx of foreign talents further caused discomfort among the local students who were grappling with the deprivation of local university places and depressed wages at the lower end of the scale.[14]

[12] See Hui and Helfer (2016).

[13] For an analysis of the 2011 General Election results, see Ortmann (2011) and Chiang (2015).

[14] In 1998, the Committee on Singapore's Competitiveness (CSC) called for an increase in foreign talent, arguing that an inflow of foreign talents would not result in "a loss of jobs for locals, but will instead increase the economic pie for all to benefit". Led by Lee Yock Suan, the Committee recommended "setting aside more places for foreign students in our education system from primary to tertiary levels and by offering more scholarships to foreign students to study in Singapore".

But immigrants continue to play an important role. Highly skilled migrants, wrote Hui Weng Tat and Cindy Helfer, "have provided Singapore with knowledge and expertise, essential to the shift in Singapore's economy from labor-intensive to capital-intensive industries and enhancing economic growth through higher productivity and innovation. The ready availability of labour augmented by the inflow of migrants in Singapore has minimized the risk of labour shortages, thereby helping to attract foreign investments".[15] In a speech delivered at the forum organised by the Economic Society of Singapore, Deputy Prime Minister Lawrence Wong noted that Singapore must remain open to top talent from around the world to "push the frontiers of possibility" while outlining the importance of developing its own people to thrive. If policies are too restrictive, Wong warned, "Our economy will contract, incomes will decline and go into a tailspin. And we will end up with far worse problems and, ultimately, it is Singaporeans who will pay the price".[16]

The decision to migrate is not to be taken lightly as it involves monetary and physiological costs. One reason to migrate is to seek a good job. There are individuals who will look for a job and settle at the place where the job is presented. But increasingly, more college-educated millennials choose where to locate first, and only then will they start to look for a job.[17] Younger and educated persons are attracted to places that offer exciting street life where they can work, shop, eat, learn and play. They would want to leave the place of work and walk or cycle to a grocery store to shop or a restaurant to get something to eat, seeing travelling in a motorised vehicle as unnecessary. The savings are spent elsewhere and, in most cases, retained in the local economy. Glen Pape puts this across succinctly, "They (the millennials) are now the largest generation and an indicator for placemaking. Their preferences for walkability and activity centers and lack of interest in driving draw them to communities with that cluster of amenities. Recent market preference studies show a clear preference for mixed-use walkable places with options for social activity".[18] Walkability matters. It attracts talent.

[15] Hui and Helfer (2016, p. 227).

[16] Goh (2002, October 18).

[17] Frumkin et al. (2004).

[18] Pape (2014).

Singapore endeavours to be the home for the best talent. Singapore is home to internationally renowned universities such as the National University of Singapore (NUS) and Nanyang Technological University (NTU) which produce globally competent local, regional and global talent but that in itself does not guarantee that the talent would remain in Singapore after graduation. Traditionally, Singapore has positioned itself as an attractive place to stay because of its human capital base, its education system, which is well regarded, its safety and almost crime-free environment, its fiscal incentives, and its population with a strong global outlook and bilingual in English and an Asian language, which taken together, offers both global and local companies access to regional and international talent.

The focus today is to create a vibrant place in the city-state, incorporating places to play, work and live. The 200-hectare one-north hub, located at Buona Vista, is a case in point. One-north has been designed to attract regional and global companies, and to help them recruit talents in Asia and elsewhere. It attracts talents and their families by offering places with high level of walkability, housing choice and activities that will draw millennials.

The development of One-north can be traced to the 1991 Concept Plan, which set the ground for the creation of the technology corridor to attract knowledge and innovation-driven industries and provide researchers with an integrated place to work, live and play. One-north, a name made with reference to Singapore's geographic location of one-degree north of the equator, offers a unique urban design, home to large corporations such as Essec Business School, Insead Asia Campus, GlaxoSmithKline, Lucasfilm, Procter and Gamble, Seagate and local institutions like MediaCorp, Singapore Polytechnic, National University Hospital and Agency for Science, Technology and Research (A*Star). One-north consists of three distinct buildings, each dedicated to a specific cluster. They are the Biopolis—the biomedical science hub with over 260,000 square metres of space to promote Singapore as the location of biochemical research and development and manufacturing; Fusionpolis—comprising 120,000 square metres of space to integrate science, engineering and other disciplines and Mediapolis—which serves as the home for digital production and broadcast facilities.[19]

[19] Clustering is one the strategic approaches of the Singapore government to push the production frontier outward. EDB is largely responsible for reaching out to corporations

One-north adopts the mixed-use approach with leisure, housing and green spaces or more generally as a self-contained precinct to create opportunities for informal interaction and to foster cross-fertilisation of ideas to spur innovation and new ideas. Buildings are located close to one another. The ground levels were lined up with food and beverage outlets to encourage and facilitate interaction. The height of kerbs is reduced to lower the obstacles when pedestrian steps onto the road thereby facilitating the ease of movement throughout one-north. Pedestrians are sheltered and shaded, and the slightly curved roads help to funnel air to provide a constant breeze to make it more comfortable for pedestrians. Many of the flora and fauna, including the mature trees, were retained in one-north, with hidden pipes running up the trucks to sprinkle water from above. Sky bridges are constructed to connect buildings and the various institutes. To create a safe and walkable environment for pedestrians, traffic calming approaches using hones granite cobblestones are used to pave internal roads instead of the asphalt.

Parks, shopping malls and neighbourhood centres are in adjacent neighbourhoods to enhance the residents everyday setting and experiences. Street level activities are possible as the building and road designs allow for open blocks of various sizes to be created. High tech companies are attracted to One-north for its proximity to business and non-business activities. This is especially beneficial to smaller companies who do not have to offer amenities like eating places or health facilities for their employees as these are readily available in the immediate neighbourhood.

Communication and work meetings at One-north do not have to be confined to within the buildings. They can take place under shelters and trees. The clusters are located close to institutions of higher learning, presenting ample opportunities for academic-industry partnerships, and sharing. Surrounding the clusters is the suburban infrastructure, including the public housing, which itself is self-contained, comprising a wide range of facilities to help residents lead a healthy and active lifestyle.

One-north is designated as a little bohemian in Singapore, like Holland Village, Siglap and Clun Street where creativity and innovation could

all over the world and bring them to Singapore to setup their business. JTC is the statutory board in-charge of allocating industrial land and space in the city-state by supplying various types of industrial facilities including prepared industrial land, standard factories, flatted factories, business parks and warehousing complexes to serve the industrialists' needs.

strive and flourish. It represents a part of Singapore's effort to become a culturally vibrant city to attract global talent who prefer diversity, openness and an authentic neighbourhood environment.[20] One-north is positioned as a place which promotes community building, organising events throughout the year for businesses and their associates to network and connect. At Fusionpolies, for example, sports facilities and a 650-seat foot hall were added to create an environment for people to unwind, network and meet-up. By creating such an environment in one-north, the plan was to attract the best and brightest around the world to live and work in Singapore.

There are numerous choices when it comes to where to live at One-North, including black and white colonial era houses and apartments in Wessex precinct, which have been converted to home offices and home studios to foster a creative community. The vicinity is surrounded by trees and greeneries, a great choice for nature lovers. There are also housing estates in Dover and Commonwealth which are equipped with reputable private and public schools. Condominiums and hotels which are suitable for shorter-term stayers are located within walking distance from one-north. In 2002, JTC took over the management of Chip Bee Gardens, a residential area in Holland Village (located close to one-north) to provide another residential option in one-north.

The development of Mediapolis is worth further mentioning. It represents a significant departure from Singapore early industrialisation strategy which focused on science and technology. To evolve as a global city with total business capabilities, Singapore has embarked on the journey to become a Renaissance City as early as the mid-1980s following its first recession since independence to cultivate the creative class and bring about diversification in the portfolio of industries. Singapore strives to be the location for value-creating collaborative work in digital contents, visual effects, content hosting and be part of the growing industry which integrates technology and creative content creation. A number of initiatives were introduced, including setting up the LaSalle College of the Arts, National University of Singapore (NUS)'s Yong Siew Toh Conservatory of Music and Nanyang Technological University (NTU)'s School of Art, Design and Media, which became the first local higher education institution to offer the Bachelor of Fine Arts degrees. In 2021,

[20] Florida's (2002b) study on innovation in the United States found strong correlation between the bohemian index and high-technology concentration.

the government made an announcement about the establishment of the University of the Arts Singapore (UAS), which is set to become the country's first and only arts university. The aim is to cultivate a new generation of visionary artists, designers, educators, researchers, arts leaders and changemakers. UAS is formed through a collaboration between two esteemed arts institutions in Singapore, namely LASALLE College of the Arts (LASALLE) and Nanyang Academy of Fine Arts (NAFA). World class musicals such as Cats and The Phantom of the Opera were brought in to further enhance the attractiveness of Singapore as a Media City, Tourism Capital and Cultural Capital, and reinforce the aspiration of locals and foreign talent where people can work, play and live a more well-rounded and enriching life. Media professionals appreciate human diversity. They are fine with busy and fast paced lifestyle. But they want to work in places which offer a variety of cultural events and different options in food and arts scene. They would prefer to get to places quickly and avoid long commutes to get from one place to another.

In sum, for the creative class, having in close proximity the places to eat, day-care centres to look after the kids while they work and green spaces to relax after a hard day's work are important considerations in determining where to work and locate. Density, mixed-use and walkability—features of one-north—are all keys to attracting talent in this sector.

Encouraging Shopping

The approach to urban design and its impact on the economy goes beyond talent attraction and retention. A pleasant walkable climate shapes our shopping habits and the performance of the services sector. Keynesian economics tells us that an increase in consumption by a dollar leads to more than a dollar increase in gross domestic product. The phenomenon is known as the multiplier effect. Encouraging consumption is one way to increase the country's national income and lift the economy out of recession. Getting the people out to shop and make purchases makes economic logic to drive economic growth and create jobs. Indeed, shopping is a common consumption activity, and forms part of the lifestyle of the population. Shops selling groceries or providing essential services such as haircut, or shoe repair contribute to the economy's national output and provide employment opportunities. Their survival is therefore essential, which depends on a host of factors, including location and travelling option vicinity which houses them.

The choice of shops is based on the shopper's willingness to travel for the different types of goods and services; people generally are more willing to travel by foot or bicycle for casual purchases but will usually travel further by bus and cars for major purchases such as clothes and electrical appliances.

The Singapore government is actively promoting the use of walking, cycling and other alternative modes of transportation for shopping purposes. This proactive approach aims to create a sustainable and vibrant economy while encouraging people to adopt healthier and eco-friendly means of getting around. The involves providing more choices and better-quality public transportation, including linkways to shopping centres and public transport stations and provision of air-conditioned buses and bus shelters to enhance of the comfort and ease of commuting by public transportation. Comparatively speaking, fares, which are controlled by the Public Transport Council, continue to be affordable. The use of shared value fare payment scheme in the form of magnetic cards and contactless smart cards further enhanced the ease of using public transport. Passengers are able to time their daily journeys and minimise waiting time for bus services by referring to the time schedule in the bus stops. Integration of transportation services is common in Singapore where bus stations, LRT and MRT stations are located within walking distance to encourage commuters to transfer from one form of public transportation to another.

Muhammad Faishal Ibrahim collected data from 675 respondents in Singapore and considered their perception of the various modes of transportation for shopping, considering attributes such as comfort, safety, time and cost. They found that walking and cycling are strongly perceived to be more suitable for short-distanced travel. Perhaps unsurprisingly, comfort is commonly perceived among the respondents for travelling by car and taxi. Faishal wrote, "…walking for shopping purposes recorded reasonably good perception ratings in Singapore. In fact, it recorded higher overall weighted factors rating than the bus. It has relatively higher ratings for its ease of travel, cost, safety, suitability and directness for shopping purposes".[21] Walking is deemed as the easiest way to travel.

[21] Ibrahim (2005).

Any improvements in other factors like safety and comfort of walking will likely encourage more walking, even among car owners, to shopping centres which are within walking distances from the workplace or home. Some of the measures include providing "covered walkways, brightly lit walking routes, construction of ramps and attractive landscapes to make it easy and pleasant for shoppers who walk for their shopping trips". This would be good for business and the economy. On 30 September 2022, the Straits Times ran a story to depict the concerns over pedestrian safety at the Bedok Mall carpark entrance in the eastern part of Singapore. Cars entering and exiting the mall might not see the pedestrians crossing in front of the entrance. The concern was brought to the attention of the Kampung Chai Chee ward in East Coast Group Representation Constituency (GRC), Tan Kiat How, who took action to deal with the problem, including installing a bump at the carpark exit to slow down vehicles exiting from the mall, adding prominent signs to warn motorists of pedestrians crossing ahead, installing convex mirrors to help pedestrians see vehicles exiting from the mall and installing signs like "caution" and "look out for incoming vehicles" to alert pedestrians when they cross the carpark entrance point.

Shopping in the central district is concentrated in Orchard Road. The name was acquired as the area housed many gambier, pepper, nutmeg and fruit plantations in the 1830s. The shopping belt was developed in the 1950s. During the next decade the stretch of road became the prime shopping and hotel district, home to Robinson's, CK Tang, Metro and many other household names. The early 1970s saw the completion of shopping centres such as Shaw Centre and Far East Shopping Centre along Orchard Road further confirming the area as the place to go for shopping. With its length of more than 2 kilometres, Orchard Road is regarded as one of the greatest shopping streets in the world. Goh Hup Chor and Heng Chye Kiang told us, Orchard Road is today "a five-lane arterial boulevard with commercial developments on both sides creating a continuous shopping and entertainment belt". Orchard Road "has a well-defined streetscape unified by roadside tree plantings and wide pedestrian promenades" which is part of the overall design, which also includes "guiding principles for building setback, public space provision, signage, and lighting". In terms of building setback, "a distance of 7.6 to 11.6 metres from the road reserve is constantly applied to determine the building line for both sides of Orchard Road in order to create the effect of a grand boulevard".[22]

With rising number of tourists and growing affluence of the local population, Orchard Road is very popular with the shopping malls enjoying high occupancy rates. The stretch of road is easily accessible by foot, enabling pedestrians to visit one shopping centre after another within proximity. However, pedestrians are not allowed to cross the road at-grade intersections. They had to use underpasses thereby posing some inconveniences to the shoppers. This continues till this day. As Bruno Wildermuth wrote, "in many parts of Singapore's central area" including the intersection of Paterson Road and Orchard Road, "pedestrians are required to cross via a three-legged path rather than simply straight across, presumably to speed up the turning car traffic". This is unlike Swiss cities, Wildermuth noted, where pedestrians "have equal rights to the use of the road space and many Swiss cities have areas where motor vehicles are generally banned except for delivery services". Citing the case of Bahnhofstrasse, a shopping street like Orchard Road which is reserved for trams and pedestrians only.[23]

The vicinity has since been bolstered up as far as shopping in the Orchard Road is concerned with underground shopping. The completion of Suntec City, Marina Square and Millennial Walk in the 1980s and 1990s marked an extension of the central district to cover the Marina Centre. In conjunction with the development of the MRT system and an effort to reduce traffic problems, underground shopping malls have been developed, and travellators are installed to allow shoppers to move from one shopping centre to another comfortably on foot. The street level entrance to the ION Orchard, for example, is also the entrance to the Orchard Road MRT station, taking the form of a huge glass cocoon.[24] The concept has since caught on in other shopping districts to link shopping centres which are in close proximity. The changing system of directing the movement of shoppers propels the importance of the

[22] Goh and Heng (2017, p. 217).

[23] Wildermuth (2016, p. 109).

[24] ION Orchard, one of the largest malls in Orchard, offers the experience of walking through a 'spatial light concourse' and 'the only immersed or "surround" LED walls' which flank the escalators. The LED walls are programmable to suit the time of day and festive occasions (Hudson, 2015).

Orchard Road area as a primary shopping district in Singapore, generating jobs and income to the population.

In his 2005 National Day Rally speech, Prime Minister Lee Hsien Loong announced renewed effort to revitalise the shopping district. Lee told Singaporeans:

> In the city centre, we will rejuvenate Orchard Road. It's one of our premier shopping districts, a brand name, known all over the world but it's facing competition from shopping districts in other cities; Nanjing Lu Road. We've already done some things, we've opened up. You've now got people, food, cafes on the sidewalks. We've got the malls opened up, so from the street you can see what's going on. From inside you can watch the passing crowd. We have a vibrant street life—dances, drummers, entertainers. But we can do more. We'll get the owners of the malls to do more and if you look at the beginning of Orchard Road right at the top of Orchard Road, there's a vacant picnic site over the Orchard MRT Station which is very popular, but I think it's a prime site and we'd like somebody to develop it, a new focal point with space for events and an observation tower. And we'll make Orchard Road one of the great streets in the world, a place to see and be seen, a place for all to enjoy ... It's very interesting what's happening. We're transforming it [the area around Bugis], bringing life back to it, activities, making it very exciting.[25]

Today, the Marina Centre is one of the most prominent sites to represent the past, present and future. It houses the Integrated Resort and stages national event such as the annual National Day parade and global events such as Formula One (F1) race and the Summer Youth Olympics. Besides Orchard Road and Marina Centre, the central area consists of key physical landscape of the Singapore River. The clean-up campaign of the Singapore River, an initiative by Prime Minister Lee Kuan Yew in 1977, took about 10 years to complete. Today, the Singapore River is divided into three zones—Boat Quay, Clarke Quay and Robertson Quay—each with its unique set of pedestrian-friendly design, catering to commercial activities, including shopping, adding jobs and tourists. Clarke Quay, for example, houses a cluster of shophouses, restaurants and entertainment outlets with pedestrianised lanes. Then, there are the New Bridge-Victoria Street corridor, and areas surrounding Kampung Glam and Little India

[25] Quoted in Hudson (2015, pp. 289–290).

where street blocks of traditional houses are present and easily assessable by foot due to their proximity to the MRT stations.

Singapore's national brand assumes a major part in motivating tourists and residents in engaging activities such as walking and cycling as well as enabling tourist walkability. An ambience of rich Asian culture and heritage, safety and comfort awaits tourists and residents, offering a high level of comfort for tourists and residents due to innovative technologies, convenience of transport interchanges, well trained professional staff in the service and entertainment sectors. A brand popularised by iconic sites such as Orchard Road and Singapore River is a combined representation of the socio-cultural traits and natural resources of a nation with an after-effect of the nation's political and economic situations. For tourists who desire for a short getaway, Singapore is one of the most preferred destinations because one can visit all the key city attractions in a short period of time. Notably, Singapore's national brand has much to offer from its historical times to the modern times and something technological and futuristic which trigger constant interest from residents and locals.

There have been calls for years to pedestrianise Orchard Road to allow for more street level events. While recognising the difficulty to remove cars completely, many see the possibility to offer an urban experience with diversity of attractions to attract diverse communities. Steps have been made towards this direction. The Somerset vicinity now dedicates space for youth and entertainment whereas Dhoby Ghaut has added greenery filled zones with family friendly attractions. URA announced in September 2022 the plans to pedestrianise a stretch of road from Buyong Road to Handy Road to allow for street level activities and three hectares of the Istana Park to be expanded to about thrice its current size as part of the government's plan unveiled in 2020 to transform Orchard Road into a "lush green corridor".

In the early years, shopping in Singapore was largely at the central of the city.[26] Exceptions are found, for example, the Beauty World along Bukit Timah Road and Holland Village Shopping Centre in Holland Village, which is the home of a department store, supermarket and handicraft and antique selling shops, catering mainly to economically more well-off locals and expatriates. Decentralisation has led to extensive growth of small-scale retailing businesses. Small scale retailing activities

[26] For details, see Dale (1999).

in fixed shops dominated in the new towns and public housing, leading to a proliferation of small family run businesses, taking over hawking and street vending in the bazaar economy prior to the 1970s. The decision to decentralise in the 1970s and 1980s was to take the pressure off shopping in the central business district and address rising population and congestion in the central area. New towns were created, housing a wide range of goods and services providers such as provision shops, wet markets and others, offering jobs to the locals to earn a living. They form a part of the small and medium sized enterprises in Singapore which makes up more than 95% of total enterprises, employing more than 70% of workers and contributing more than half of the value added generated by all enterprises.

Muhammad Faishal Ibrahim and Lim Fong Peng have classified shopping in new town areas into three types; traditional neighbourhood centres (TNC), new generation neighbourhood complexes (NGNC) and neighbourhood cluster shops (NCS).[27] The TNC was established in 1970s to provide shopping and other services for residents in the new towns. The intent was for residents to gain access to essential goods and services within short convenient walking distances from the surrounding housing blocks. TNCs are evidenced in older housing estates such as Toa Payoh, Ang Mo Kio, Bedok and Queenstown. The TNCs are crowded, not surprisingly in view of the extensive offering of services, which leads to problems such as noise, heat, fume and nuisance arising from the business operations; tolerable for the older generation of shoppers but appears to be less favourable for the younger shoppers.

The NGNC was established in 1991. They are located within a five-mile radius from the residential flats, offering an alternative to suburban population to do their shopping. The air-conditioned complexes of size of between 10,000 square metres and 40,000 square metres and between 100 and 200 shops are designed like a typical shopping centre, accompanied by covered parking area, fast food restaurants, convenient stores, clinic, electrical stores and others. They provide a clean and landscaped shopping environment, and, as Ibrahim and Lim put it, "a one-stop shopping experience to residents". Unlike the TNCs, which are located closer to the residents, the NGNCs are lesser in numbers and meant to cover a larger number of residents. This translates to longer travel distances

[27] Ibrahim and Lim (2005).

for residents further away, without the luxury of being within walking distance to the shops. Residents would need to take a bus or the LRT to reach the complex.

The NCS houses a limited range of serves, typically a convenient store, clinic, saloon and an eating house. They serve a small local community, locating below or near multistorey carparks within the housing estates. Most are within walking distance to residents around them, providing the convenience to residents to do their shopping. Proximity of the neighbourhood centres is a key factor in affecting the choice of shopping experience. Faishal and Lim's study, which was based on interviews, showed that on average, 54% of the respondents get to the neighbourhood centre by walking and 24% by bus. The travelling time from their home to the neighbourhood centres ranges between 5 and 20 minutes. There was a strong preference for TNC based on shorter travel time, lower cost of travel. The fact that TNCs offer a wide range of services—availability of wet markets, eating establishments and a wide variety of merchandise—also matters—contributing to shoppers' preference for TNC as compared to the NGNC despite the latter's higher standards in cleanliness, security and parking facilities.

Many cities have realised the importance of ensuring good walkability. Getting it right, walkability creates liveability cities which people want and in turn bring in people to boost the city's competitiveness. As such, cities around the world have announced plans to adopt people-centred approach in urban design and development by making parts of the cities more pedestrian friendly. Innovative urban redevelopment projects like the High Line project in New York involves strong public and private sector participation. Adjacent streets could take advantage of the development to maximise the land-use potential. In High Line project, to take an example, the redevelopment initiative has spurred the development of a slew of commercial and residential projects which are well connected. Projects such as High Line, Brooklyn Bridge Park and New York City Plaza has made New York City an attractive to work and live.[28]

In this chapter, we show that walkability contributes to the economy and aligns with the economy first mentality in Singapore. As part of the decentralisation effort, Singapore has developed commercial nodes to shift commercial activities and jobs to other areas and reduce traffic

[28] See Cheong (2019).

congestion in the city centre. Commercial regional hubs in Tampines, Jurong East, Paya Lebar and Novena are well-designed, consisting public housing, amenities, premises for businesses and well connected through high-speed rail and highways. Even if one does not live in the area would have visited the neighbourhood towns to visit the malls or use some of the amenities in the town like the market and hawker centre. Consider the case of Jurong Lake District (JCD). JCD has been envisioned as the second Central Business District in Singapore. Located in the western part of Singapore, JCD offers exciting opportunities to develop a car-lite district with located within walking distance. A fully pedestrianised zone is created around the Jurong East Interchange, connecting shopping malls and other amenities. Big trees are planted to provide sheltered pathways to facilitate walking and cycling. The commercial hubs bring economic growth opportunities for Singapore. That urban design is targeted to facilitate activities and movement of people to support economic planning is illustrated. As Ho Kong Chong told us, "...urban planning has to be related to the economic planning in city building, with economic planning providing the broad strokes in shaping the Singapore economy and urban planning in facilitating and housing the activities that are the result of economic planning".[29]

References

Abin, R. (1991). Developments in Indonesia-Singapore bilateral relations: Politics. In T. Lau & B. Singh (Eds.), *Singapore-Indonesia relations: Problems and prospects*. Singapore Institute of International Affairs.

Birch, D. (1994). Cultural studies in the Asia Pacific. *Southeast Asian Journal of Social Science, 22*(1), 1–12.

Cheong, K. H. (2019). *Seeking a better urban future*. World Scientific Publishing Company.

Chiang, H. D. (2015). *Elections in Singapore*. Straits Times Press.

Clarke, G., & Monk, A. (2010). Investment Corporation (GIC): Insurer of last resort and bulwark of nation-state legitimacy. *The Pacific Review, 23*(4), 429–451.

Dale, O. L. (1999). *Urban planning in Singapore: The transformation of a city*. Oxford University Press.

Florida, R. (2002a). Bohemia and economic geography. *Journal of Economic Geography, 2*, 55–71.

[29] Ho (2017, p. 273).

Florida, R. (2002b). *The rise of the creative class*. Basic Books.
Frumkin, H., Frank, L., & Jackson, R.·(2004). *Urban sprawl and public health: Designing, planning, and building for health communities*. Island Press.
Goh, D. (2001). The politics of the environment in Singapore? Lessons from a 'strange' case. *Asian Journal of Social Science, 29*(1), 9–34.
Goh, H. C., & Heng, C. K. (2017). Shaping Singapore's cityscape through urban design. In C. K. Heng (Ed.), *50 years of urban planning in Singapore*. World Scientific Publishing Corporation.
Goh, Y. H. (2002, October 18). To thrive, Singapore must stay open while developing own talent: DPM Wong. *The Straits Times*.
Ho, K. C. (2017). Era of globalization: Singapore's new urban economy and the rise of a World Asian City. In C. K. Heng (Eds.), *50 years of urban planning in Singapore*. World Scientific Publishing Corporation.
Hudson, C. (2015). ION Orchard: Atmosphere and consumption in Singapore. *Virtual Communication, 14*(3), 289–308.
Hui, W. T., & Helfer, C. (2016). Migration policies: Lessons from the Singaporean and Swiss experiences. In Y. Guo & J. J. Woo (Eds.), *Singapore and Switzerland: Secrets to small state success*. World Scientific Publishing Corporation.
Ibrahim, F. (2005). Attitudes to transport modes for shopping purposes in Singapore. *Transport Reviews, 25*(2), 221–243.
Ibrahim, F., & Lim, F. P. (2005). The development of neighborhood centers in Singapore: From traditional to cluster shopping. *Pacific Rim Property Research Journal, 11*(4), 373–392.
Jacobs, J. (1972). *The death and life of great American cities*. Penguin.
Lye, L. H. (2008, July). A fine city in a garden—Environmental law and governance in Singapore. *Singapore Journal of Legal Studies*, 68–117.
Ortmann, S. (2011, November). *The 2011 elections in Singapore: The emergence of a competitive authoritarian regime* (Working Paper No. 112). City University of Hong Kong.
Pape, G. (2014). *To attract talent think what kind of place the emerging generation of professionals want to live*. https://www.canr.msu.edu/news/to_attract_talent_think_what_kind_of_place_the_emerging_generation_of_profe. Accessed 10 June 2022.
Quah, J. (2008). Good governance, accountability and administrative reform in Singapore. *American Journal of Chinese Studies, 15*(1), 17–34.
Singapore Management University. (2019, March 25). University builds bridges to innovation. *The Straits Times*. https://news.smu.edu.sg/sites/news.smu.edu.sg/files/wwwsmu/news_room/smu_in_the_news/2019/Mar2019/Mar25/20190325-STHM-Home-B06-33x41.pdf. Accessed on 7 April 2023.

Tan, E. C. (2020, September 10). *Universities need to tear down subject silos.* https://www.straitstimes.com/opinion/universities-need-to-tear-down-subject-silos. Accessed on 7 April 2023.

United Nations Conference on Environment and Development. (1992). *Singapore's national report for the 1992 UN Conference on Environment and Development Preparatory Committee.* Inter-Ministry Committee for the UNCED Preparatory Committee.

Wildermuth, B. (2016). Land transport policy: Urban infrastructure in Singapore and Switzerland. In Y. Guo & J. J. Woo (Eds.), *Singapore and Switzerland: Secrets to Small state success*. World Scientific Publishing Corporation.

CHAPTER 9

Smart Nation and Walkability in Singapore

It is well known that advances in Information and Communication Technology (ICT) such as cloud computing and sensor networking have allowed for collection, processing and application of vast amounts of data and information to monitor and control the urban systems. Web-based tools are used to gather information about street conditions and their level of support for walking and cycling. Walking audits consider and promote walkability as a viable form of transport. They are usually undertaken by a range of different stakeholders including local community groups, transport planners, urban planners and politicians. Walk audits or walking assessment tools are traditionally paper based with the auditors documenting barriers and physical features while taking a walk with community members to assess street infrastructure and conditions. Sophisticated free online geospatial services like Google Steet View, Google Earth, Google Maps, Google Earth Pro and Microsoft's Bing Maps and Streetside are now increasingly used to make places more walkable.[1] The rise of autonomous vehicles, digitalisation and real-time information facilitates monitoring of traffic movement. The use of smartphones to hire a private car or order items for delivery to homes and

[1] See Brookfield and Tilley (2016), Rundle et al. (2011), Ben-Joseph et al. (2013), Clarke et al. (2010) and Schlossberg et al. (2015).

Melissa Liow Li Sa and Sam Choon-Yin, *Sustainable Urban Development in Singapore*, Sustainable Development Goals Series,
https://doi.org/10.1007/978-981-99-5451-3_9

offices reduce the use of cars and alleviate road congestion, potentially freeing up road and parking space which could be converted to pedestrian walkways or bicycle parking space.

Smart City provides a useful theoretical framework particularly in the case of Singapore. The theory proposes that ICT can be effectively used to create more liveable and sustainable cities by improving the efficiency and effectiveness of urban services, such as transportation and public utilities. In the context of walkability, ICT can be used to enhance pedestrian experiences and increase the safety and comfort of walking, for example, by providing real-time information about traffic patterns and pedestrian routes. Singapore has also explored ways to allow people to interact with technology and how technology can be designed to enhance human experiences. Specifically, the Human-Computer integration approach has enabled urban and transport planners in Singapore to design and develop ICT tools that support and encourage walking, such as pedestrian navigation systems, health and fitness tracking apps and augmented reality systems that enhance the experience of walking. Through an analysis of the Smart Nation initiative, this chapter documents Singapore's attempt to integrate the city with smart technology to overcome its limited resources and promote walkability.

Singapore's IT Journey

Singapore's technology journey began in 1980. The government saw technology as a way to improve efficiency and recognised the necessity for the government to take the lead. At the same time, the government knew that the government could not do this alone and would need to establish a partnership with the private sector. To promote the tech-focused vision, Singapore welcomes local and foreign companies' participation in new product testing and development, often with financial support from the government.

In 1980, the government formed the Committee on National Computerisation (CNC). CNC was chaired by Tony Tan Keng Yam who had led OCBC's computerisation efforts before he entered politics in 1979. He was joined by Philip Yeo, Tan Chin Nam, Lim Swee Say and Teo Ming Kian who represented the early batch of government scholars who were trained in computer science and related subjects, to develop Singapore computer policies and strategies. CNC proposed setting up of the National Computer Board (NCB) to promote the use of information technology (IT) in Singapore. Established in 1981, the NCB was responsible for the development of IT policies and plans, as well as the implementation of IT projects and programmes. The government was wholeheartedly committed to embrace information technology. The National IT Plan (NITP) was launched in 1986 to promote the use of IT in Singapore's economy and society. The NITP aimed to increase the productivity and competitiveness of Singapore's industries through the use of IT. The NITP also aimed to promote the development of IT infrastructure and human resources. This was followed by the 1992 report *A vision of an intelligent island: the IT 2000* report published by NCB. The report outlined Singapore's plan to build state of the art computing, optical-fibre transmission and cybernetic-feedback-systems capable of integrating homes, workplaces and public institutions. The IT2000 report encapsulated the future of IT in these words:

> In our vision, some 15 years from now, Singapore, the Intelligent Island, will be among the first countries in the world with an advanced nation-wide information infrastructure. It will interconnect computers in virtually every home, office, school, and factory.[2]

Renowned for its efficient and decisive approach, Singapore demonstrates a remarkable ability to swiftly translate decisions into action. With a strong focus on long-term policy planning, Singapore wastes no time in initiating the necessary steps to bring about positive change. The late 1990s saw the convergence of information technology and telecommunications, which led to the development of the e-Government Action Plan (2000–2003), which focused on rolling out government online services. The e-Government Action Plan was succeeded by the i-Government 2010 Master Plan (2006–2010). SingaporeOne was launched in 1998 to link all users, including business and government agencies on a fast broadband network, facilitating the delivery of video and media contents speedily and smoothly to homes. In the same year. the electronic road pricing (ERP) system was introduced. The system makes the motorist 'pay as you go' every time the vehicle passes the gantry post along the designated roads. The electronic scanner automatically deducts an appropriate fee from a cash card installed in the vehicle, replacing the paper-based

[2] National Computer Board (1992, p. 10).

passes and hundreds of traffic wardens stationed at the entry points to check the display of the passes. The National Infocomm Competency Framework (NICF) was launched in 2003 to promote the development of infocomm (information and communication technology) skills in Singapore. The NICF aimed to improve the competitiveness of Singapore's workforce through the development of infocomm skills and knowledge. Today, almost every government service is provided online.

NCB and the Telecommunication Authority of Singapore merged in December 1999 to form the InfoComm Development Authority of Singapore (IDA).[3] The IDA was responsible for the development of infocomm policies and plans, as well as the implementation of infocomm projects and programmes. The agency published the *Intelligent Nation 2015*, a sequel to the IT 2000 report. Like its predecessor, the report provided a holistic approach (the iN2015 plan) to promote the development of an intelligent nation through the use of IT and infocomm technologies. The iN2015 plan aimed to transform Singapore into a knowledge-based economy and society through the development of IT infrastructure, human resources and applications. Wireless@SG project was launched in 2006 to provide Wi-Fi hotspots around the island, providing users free access to the Internet in popular places on the go, including cafes and libraries. The initiative contributed to the growth of smartphone users and mobile phone subscriptions, laying out the foundation for more ambitious plans i.e., smart services. More than 90% of households today are Internet enabled and has a very high mobile phone penetration rate (on average 1 person to 1.5 mobile phones).[4]

The Infocomm Media Development Authority (IMDA) was established in 2016, as a merger between the IDA and the Media Development Authority (MDA), to lead the development of the infocomm and media sectors in Singapore. The IMDA aims to develop a vibrant and inclusive digital economy and society, by developing the infocomm and media sectors and promoting the use of IT and infocomm technologies in all sectors of the economy. The Infocomm Media 2025 report (IM2025) was significant as it outlines Singapore's aspiration to become the world's first Smart Nation. Released in August 2015 and developed by IDA, MDA

[3] Arun and Yap (2000) outline the IT arrangements at home, at work, on the road, in public service and in the court room.

[4] As reported by the World Bank. See https://data.worldbank.org/indicator/IT.NET.USER.ZS?locations=SG. Accessed 20 July 2022.

and the Ministry of Communication and Information (MCI), IM2025 frames technology as the key enabler and integral to Singapore's next phase of development, building on earlier investments in connectivity and Internet penetration.

The Smart Nation programme was launched in 2014 to promote the development of a smart nation through the use of IT and infocomm technologies. The programme aims to create a more connected and efficient society through the use of technology in areas such as transportation, healthcare and urban planning. Strong emphasis has been placed on upgrading the universities and research centres, for example, through the creation of the A*STAR (2001), the Research, Innovation and Enterprise Council (2006), and the National Research Council (2006). Singapore has also supported strategic partnerships with globally renowned science research institutions like the Massachusetts Institute of Technology from the United States, Cambridge University in Britain, Peking University and Shanghai Jiao Tong University from China, Technical University of Munich from Germany and others to exploit linkages and generate ideas to technical significance. At the launch of the Smart Nation initiative on 24 November 2014, Prime Minister Lee Hsien Loong envisaged Singapore as "a nation where people live meaningful and fulfilled lives, enabled seamlessly by technology, offering exciting opportunities for all".[5] A key initiative was the setup of the National AI Office in 2018, to lead the development and implementation of artificial intelligence (AI) strategies in Singapore. The office aims to promote the development and adoption of AI in areas such as healthcare, transportation and finance, to help Singapore to become a leading AI hub in the region. The Smart Nation and Digital Government Group (SNDGG) was established in 2020, to help Singapore to become a leading smart nation and digital government. The group aims to develop and implement strategies for the use of technology in areas such as transportation, healthcare and urban planning, as well as to promote the development and adoption of technology in the public sector. In 2021, Singapore has announced the launch of its 5G networks, which will be available to the public in 2022. Singapore aims to use 5G to support programmes in a wide range of industries and sectors to spearhead Singapore's next phase of economic development.

[5] Speech delivered by Prime Minister Lee Hsien Loong at Smart Nation Launch on 24 November 2014. Available on https://www.pmo.gov.sg/Newsroom/transcript-prime-minister-lee-hsien-loongs-speech-smart-nation-launch-24-november. Accessed 8 July 2022.

Digitalisation of the nation requires the citizens to have the basic skillsets to use digital technologies. To enable this, the SkillsFuture initiative was launched in February 2015. Singaporeans aged 25 years old and above to given up to S$500 credit to attend courses that focus on skills training and development. A centralised website has been setup to help Singaporeans identify a wide variety of courses in information and technology, communications, accounting, finance, social services and others. For example, individuals who want to upgrade their knowledge and skills in IT can sign up for courses in Data Analytics, Coding, Digital Marketing and Digital Finance, Media and other tech-enabled services through the Government's Adapt and Grow Programme. In this regard, SkillsFuture was established as part of the Smart Nation initiative to prepare citizens for the future economy that is more IT based. Thijs Willems and Connor Graham put this across succinctly: "Central to this new economy are citizens who know how to use digital technologies and are able to assess its consequences skilfully, for instance by knowing how to reach online communities, use of specific smartphone serves, or critically assessing algorithm-based recommendations. A specific type of citizen is imagined here, and the masterplan and associated policies and initiatives such as SkillsFuture create the conditions on which citizens can be aligned with this future imagination of the nation".[6]

The outcome has been recognised. Singapore is regularly ranked as one of the most connected countries in the world. In 2014, a comparative study conducted by Accenture ranked Singapore first based on the use of "digital government"—from offering online portals to access public services to employing digital channels and social media to communicate and engage with citizens. The Network Readiness Index puts Singapore in the top 10 consistently, and the city-state takes the top spot for the Asia Pacific region, outperforming South Korea, Japan, Australia and China.

Walkability and Smart Nation

How does Singapore embrace technology as the prime driver of walkability?

Smart Nation is the key initiative to integrate ICT in the urban fabric to enrich the lives of the residents. As a Smart Nation, the focus is on

[6] Willems and Graham (2019, p. 520).

identifying and designing IT solutions in service delivery to the public, including the use of the Internet of Things, data analytics, augmented reality, robotics and others. The government collects vast amounts of data to find smart solutions to enhance the public transport system and provide greater comfort, convenience, reliability and support for Singapore's vision of becoming a car-lite society. Some of the key initiatives are summarised below[7]:

- Public buses and trains are tested with hands-free fare gates to allow the elderly, families with young children and commuters with mobility challenges to enter and exit train stations and buses quickly without tapping their fare cards on the card readers.
- Singapore introduces the Smart Nation Sensor Platform (SNSP), an integrated, nationwide platform, to collect essential data using sensors. The data is analysed to create smart solutions.
- The LTA instals sensors to collect data and information on buses' real-time location and arrival times with the purpose of improving transport planning.
- The LTA also collects anonymised data from commuters' fare cards. The purpose is to identify busy bus routes so that bus fleets can be more effectively managed based on commuter demand. LTA is seeing improvements on these fronts with 92% reduction in the number of bus services with crowding issues and three to seven-minute reduction in the average waiting time on popular bus services.
- SG Traffic Watch uses a combination of real-time data and past trends to manage congestion and other traffic conditions.
- Self-driving shuttles are tried to use technology to better connect towns and enhance mobility for commuters, particularly the elderly and persons with disabilities.

Housing about 80% of the population, Singapore's Smart Nation initiative to promote walkability must involve the public housing estates and new towns. Paul Barter had argued that the HDB housing estates reflect a "towers in the park" style of urban design with a large area

[7] Extracted from https://www.smartnation.gov.sg/about-smart-nation/sndgg. Accessed on 2 July 2022.

of space for roads, suggesting that the government has given priority to motorists and low priority to pedestrians at intersections.[8] To be fair, the HDB main priority in the early years of independence was to solve the housing shortage problem. With the problem largely solved, the HDB can turn to focusing on identity and character, and social integration. In newer town blocks, the public housing estates are served by more comprehensive cycling and pedestrian networks free of obstructions to support a walkable city.

Indeed, the adoption of technology to improve liveability of the residents is part of HDB's continual planning. In September 2014, the HDB unveiled the Smart HDB Town Framework, which adopts smart technologies in the planning, development and management of HDB towns. The Smart Hub, which serves as a central repository of information. Through data analytics, HDB seeks to understand residents' needs and preferences and make necessary changes in the design of the spaces. HDB uses computer simulations and data analytics to improve its design of HDB flats and new towns such Punggol Town. Working with other agencies such as the National Research Foundation, Singapore Land Authority and URA, HDB uses computer models to study the effect of sun, wind and noise to enable the housing planners to improve the placement and orientation of buildings, for example, to channel wind flow to produce a cooling effect and air quality of the surroundings to create a more pleasant environment for residents. Infrastructure is upgraded in response to the physical environment to facilitate active mobility.

The use of apps to plan journeys is common in the digital age. Digital maps like Google Maps and Apple Maps are extensively used by motorists to provide directions to travel from the point of origin to the destination. As an urban planning tool, they do not always provide accurate information, leading to poor decisions based on misinformation. Afterall, the maps are produced by policymakers, marketers and technologists to add, remove, or edit landmarks on the maps, not necessarily by approved or qualified drafters, and they can be easily changed without being noticed.[9] The other problem with digital maps is that they may not be the most appropriate tool to use by cyclists. The path recommended by Google

[8] Barter (2013).

[9] Google Maps, Greg Bensinger told us, includes listings of fake businesses. See Bensinger (2021).

Maps tends to allow the cyclists to reach the destination in a shortest time possible but take the cyclists through busy and dangerous roads.

For these reasons, Singapore has developed the OneMap app, the national map of Singapore with the most detailed and timely updated information. Developed by the Singapore Land Authority and downloadable in Google App and Apple Play Store, OneMap provides street level data of paths connecting to each other. Distances are showed from point of origin to amenities such as hawker centres, healthcare, community centres, bank branches, supermarkets, haircuts, retail pharmacy and optical shops to support travellers. The 'Bus Explorer' function indicates the bus services available for a particular location, their respective bus routes and the estimated arrival time, which are useful for users to plan their travels. How well the pathways are connected to public transportation nodes is important to improve walkability. Walking and cycling are notably more likely if the participants can access public transport as part of their journey.

OneMap however is not designed specifically to support walkability, unlike other providers such as AccessMap or SideWalkScore. Developed by Tasker Centre for Accessible Technology, AccessMap maps pedestrian paths according to the individual preferences. User are able to indicate their preference with regard to the maximum uphill or downhill steepness. SideWalkScore on the other hand shows the various destinations one can travel to, based on the maximum distance of walking, with the option to indicate the sidewalk width to allow for social distancing. These are remarkably absent in OneMap. However, progress is underway. In 2021, the Ridenjoy app was launched, tapping into OneMap programming to recommend routes which would allow cyclists in Singapore to use just park connectors and cycling paths thereby providing a safer and shorter route option for its users. The app has the function to inform users about other matters such as accidents on a route and an SOS function to trigger alert to the relevant authority in case of emergency.[10]

The efforts to build a safe and comfortable environment for pedestrians is understood as an effective mean to promote walkability. Using walk audit instruments including OneMap and Ridenjoy helps to identify cracks on the pavement, which may create trip hazards. OneMap and others like Google Maps can identify areas where low-speed traffic zones

[10] Toh (2021, December 19).

should be introduced, for example, roads with sharp curves and limited sightlines, to protect pedestrians against vehicles travelling at a dangerous speed. The information can be used by the relevant authority to erect road signs on pavements and roads. Comfort and safety for the pedestrians are associated with the physical conditions of the pavements. Key design elements include the width of the pavement, shelter, trees, resting place, shop fronts and other amenities should all be considered. The information can be collected by feedback and supplemented by online data sources, to take into consideration pedestrian flow and crowding. Walking distance from one point to another is another factor affecting walking behaviour. The shortest path and distance from the point of origin to the destination can be calculated by Google Street Map or Geographical Information Systems (GIS) software like ArGIS based on various walking speed. A long detour can discourage walking. By tapping into web-based imagery technologies to generate panoramic and high-resolution images of locations, the government can create a more pleasant walking experience in the public housing estates, considering, for example, the width of pathways, walkways design and destinations of places to enhance safety, comfort and convenience, which form the three walkability experiences.

In a large measure, growing integration of information communications technology in the urban fabric can enrich the lives of the residents. Smart technology can be used to encourage walking. Smartphone apps are able to send notifications to remind users to take a walk, such as every hour or after a certain period of inactivity. Singapore's Health Promotion Board has developed a number of apps, including "MyFitnessPal" and "Pedometer by MyFitnessPal" that track steps taken and distance walks, and can set daily or weekly goals to motivate users to walk more. Wearable devices have been developed to track physical activity, including steps taken, throughout the day. The use of game-like elements, like points and batches motivates people to walk more. For example, a mobile map that turns walking into a virtual scavenger hunt or a walking challenge with friends. The Singapore Health Promotion Board's "Step Up" campaign is a walking challenge with friends, where participants compete to walk the most steps in a month. The Health Promotion Board also uses social networks, such as Facebook and Instagram, to connect users and share their walking progress. Social network platforms allow users to connect with friends and share their walking progress, and offer encouragement and support. There are smartphone apps that provide an audio guide for a walking tour, giving information on historical and cultural sites while encouraging users to walk. The National Heritage Board has developed virtual reality apps and games to stimulate outdoor environments and encourage users to walk in place. Singapore has been developing its smart city infrastructure to encourage walking, including the use of sensors and cameras to conduct data on the usage of sidewalks, streets and other areas to identify areas of low pedestrian traffic and design interventions to encourage more walking. These technologies can be used individually or in combination to create a more engaging and interactive experience and encourage people to walk more and improve their physical activity levels.

It is worth noting that ICT may deter walkability as well. ICT encroaches on walking activity. Residents in the housing estates used to walk to nearby supermarkets or food centres to buy groceries and food, respectively. Today, they order these items online. More broadly, ICT revolution has created powerful corporate entities who could threaten to place self-driving cars on roads which may further force out pedestrians, much like the 1920s and 1930s when automakers pushed to build wide laned roads to support the new transportation technology, displacing pedestrians and bicycles off the streets. For ICT to productively contribute to creating a better walkability experience, the government must take the lead to understand what the users want, and not be swayed by the vision and promises of profit-making tech companies. One way is to think about progressing *around* technology, for example, by collecting and analysing accident and crime statistics, which the government could use to address safety-related issues associated with walkability (i.e., desire for freedom from hazards, including safety to crossroads and freedom from crime, including theft).

Communal Community Practice: The Case of Tengah and Punggol Digital District

In 2019, the IMD Smart Cities Index, conducted through the combined efforts of Singapore University of Technology and Design as well as Swiss Business School IMD, ranked Singapore as the smartest city in the world. The index evaluates how cities embrace digital technologies and enhance the quality of life of their residents. Key aspects of a smart city include governance and health, mobility, public safety and citizen security, digitalising employment access and upgrading community institutions. Although there is no one common description for what constitutes a smart city, it was at the height of the Internet of Things (IoT) era where smart cities harness digital technology to uplift the supply and advancement of urban services.

The Singapore Smart City invests on new project concepts in the nation's bid to reduce carbon footprint and encourage neighbourhood residents to engage in walking and cycling. City life and its hustle and bustle can be attractive but when it is overwhelmed with crowded populaces and extreme pollution, its urban areas can become some of the least environmentally friendly spaces. On this note, it is worth noting that efforts are underway to create car-free town centres and provide residents 'in-town' experiences of living sustainably and comfortably.

- Tengah

Tengah is slated to be Singapore's innovative smart city with a capacity to host 42 thousand people. Tengah is a new town in Singapore that is being developed by the HDB with a focus on green and sustainable living. It is estimated that it will provide 42,000 new homes—30,000 HDB flats and 12,000 private housing units. One aspect of this development is the incorporation of communal community spaces, such as community gardens, which are designed to promote social interaction and a sense of community among the residents. About 20% of the land in Tengah is set aside for green spaces which will be woven throughout the town, from a large central park to gardens and play areas that run through the housing estates. These spaces also serve as a way to promote environmental sustainability by encouraging the use of green spaces and reducing heat island effect. The town is designed to be car-lite, with a focus on walking, cycling and use of public transport and features a central, an event plaza, hawker centres, among other amenities. Marrying humans with nature, it is a bid in encouraging behaviours that project less carbon footprint. The public officers refer the Tengah development as a *forest town*. The naming is due to its vast spread of natural landscapes and lush greenery. The centre of the city offers an ecological corridor which links a water source to provide for the wildlife.

Tengah's layout is designed to promote walking and cycling as the main transportation means. Located within the proximity of a MRT station, there are roads for vehicles built under the town centre, permitting a dedicated space for pedestrians away from motor traffic. This gives citizens a less polluted and safer area to participate in recreational activities. Electric vehicle charging stations are positioned in various parts of the town centre to encourage energy efficient travels.

The Tengah planners are aware of the need to keep the residents cool in hot and humid Singapore but without air conditioning which accounts for up to a quarter of average Singaporean household electricity consumption. Solar powered cooling water systems, located in several sites, will be used instead of the less eco-friendly air conditioning. With places kept cool, people can walk and cycle comfortably, and relax at halfway stops designed to encourage active mobility. To further preserve energy and electricity, the Tengah developers trade for smart lights which turn on and off automatically based on the number of people in a zone at one point of time. A Pneumatic Waste Conveyance System will be constructed for trash disposal, utilising high-speed air to gather waste, reducing the dependence on garbage trucks, while generating a more hygienic and cleaner environment for citizens. A mobile app will be accessible for its inhabitants to track, report and monitor the water and electricity consumption. Getting to view utility consumption in other adjacent neighbouring blocks can encourage healthy competition among community members to bring down carbon emissions than their neighbours, nudging households to walk instead of drive and switch off lights and turn off faucets when not in use.

- Punggol Digital District

JTC Singapore is in progress to develop its first enterprise district, at Punggol Digital District. Punggol Digital District is a planned development in the Punggol area of Singapore that is being developed as a hub for the digital economy. The goal of the development is to create a vibrant, sustainable and liveable community that will attract and retain talent in the tech industry. The development will feature a range of amenities, including co-working spaces, incubation centres and a digital plaza. It will also include residential, commercial and retail spaces, as well as a variety of public spaces and parks. The development will be designed to be highly connected, with a focus on walkability and the integration of technology, such as smart city systems and 5G networks, into the infrastructure.

Punggol Digital District aims to be a model for sustainable development, with a focus on reducing carbon emissions, conserving water and energy and promoting biodiversity. The district will also feature a range of green spaces and public parks, including a waterfront promenade, to promote healthy living and a sense of community among residents. Tipped to be a car-lite district—heavy on tech but light on cars—Punggol Digital District showcases how technology can help create a more liveable and sustainable environment for the community at a local district level. Walking and cycling will take a front seat, making carbon spewing vehicles a less big deal. Eco-friendly commuting is mobilised with its seamless connected pathways, and one can cycle around on a bicycle and stroll. There are shower facilities and storage racks offered for one's convenience, and if need a ride back closer home, there are fully electric autonomous buses in service. Some of the district-wide infrastructure and services include the Open Digital Platform to integrate the management of the various buildings within a single estate system and centralise the data collection of these systems; the Centralised Logistics Hub where all goods can be dropped off and picked up thereby raising productivity and reducing traffic; the District Cooling System that centralises cooling needs; the Pneumatic Waste Collection, an underground vacuum-pipe network to eliminate the need for waste collection trucks and eradicate the odour from refuse chutes; and the Smart Energy Grid to enable consumers to adopt clean sources of energy for daily use (e.g. charging electric vehicles), and facilitate greater energy efficiency and savings (e.g. through smart metering).[11]

ICT, Walkability and Sustainable Development

Enabling walkability and incorporating smart city infrastructure and apps can help to fulfil several of the goals outlined in the Singapore Green Plan 2030 and the United Nations' 17 Sustainable Development Goals (SDGs).

One example is the Singapore Green Plan 2030's goal of creating a more livable and sustainable city by promoting sustainable transportation options, such as walking and cycling. By enhancing walkability through the use of smart city infrastructure and apps, Singapore can make it easier

[11] For details, see https://www.ura.gov.sg/Corporate/Planning/Master-Plan/Urban-Transformations/Punggol-Digital-District. Accessed on 3 August 2022.

and safer for people to walk, which can reduce the number of cars on the road, decrease air pollution and greenhouse gas emissions and improve the overall health and well-being of its citizens.

Another example is the SDG 11, which aims to make cities and human settlements inclusive, safe, resilient and sustainable. Incorporating smart city infrastructure and apps enables Singapore to improve the safety and efficiency of its transportation systems, enhance the accessibility of public spaces and services and promote sustainable urban development. The commitment to sustainable transportation is a noteworthy aspect of Singapore's efforts. Through data analysis, the government optimises public transportation routes and schedules, leading to reduced congestion, improved travel times and a cleaner environment. Enabling walkability and the use of smart city infrastructure can also help to achieve other SDGs, such as reducing poverty (SDG 1), promoting good health and well-being (SDG 3) and promoting sustainable cities and communities (SDG 11).

Overall, by enhancing walkability and incorporating smart city infrastructure and apps, Singapore can create a more livable, sustainable and equitable city, which can contribute to the achievement of several of the goals outlined in the Singapore Green Plan 2030 and the 17 Sustainable Development Goals.

In this chapter, we have discussed the use of ICT to encourage walkability. It suggests that integrating ICT with the urban fabric is a promising initiative for encouraging walking activity. This is an on-going endeavour. The development of Singapore as an intelligent island has harnessed the capacity for new technologies to emerge to support walkability. Singapore has an early advantage in ICT adoption, seeing IT as imperative to propel its economy forward.

There is evidence to suggest that a serious commitment from the government to integrate ICT in the urban fabric to enrich the lives of the residents. The improvements over the years and expectantly many more years have benefited pedestrians and cyclists, laying the path to make Singapore a leader in this area. The Singapore smart city's innovative progress is likely to heighten resident's well-being while embracing a natural landscape and reducing the nation's carbon emissions. Other nations can study from the Tengah concept, hypothetically implementing more community-based and unified methods of energy dispersal. Tengah's eco-friendly design will hopefully inspire more nations to embrace sustainable changes. On a bigger scale, contacting government officials for mandating the city dwellers to partake these sustainable changes can be advocated. A noteworthy project in Singapore that has been fully reified from similar eco-friendly concepts is the Punggol Digital District. By up-taking walking and cycling mode of transport in this cutting-edge technology district, one can get onto fully electric autonomous buses to save commuting time back home.

References

Arun, M., & Yap, M. T. (2000). Singapore: The development of an Intelligent Island and social dividends of information technology. *Urban Studies, 37*(10), 1749–1756.

Barter, P. (2013). Singapore's mobility model: Time for an update? In Institute of Mobility Research (Ed.), *Megacity mobility culture*. Springer-Verlag.

Ben-Joseph, E., Lee, J. S., Cromley, E. K., Laden, F., & Troped, P. J. (2013). Virtual and actual: Relative accuracy of on-site and web-based instruments in auditing the environment for physical activity. *Health and Place, 19*, 138–150.

Bensinger, G. (2021, September 29). How digital maps mislead us. *The Straits Times*.

Brookfield, K., & Tilley, S. (2016). Using virtual street audits to understand the walkability of older adults' route choices by gender and age. *International Journal of Environmental Research and Public Health, 13*(11), 1–12.

Clarke, P., Ailshire, R., Melendez, M., Bader, M., & Morenoff, J. (2010). Using Google Earth to conduct a neighbourhood audit: Reliability of a virtual audit instrument. *Heath and Place, 16*(6), 1224–1229.

National Computer Board. (1992) *A vision of an Intelligent Island: The IT2000 report*. National Computer Board.

Rundle, G., Bader, M., Richards, C. A., Neckerman, K. M., & Teitler, J. O. (2011). Using Google street view to audit neighbourhood environments. *American Journal of Preventive Medicine, 40*(1), 94–100.

Schlossberg, M., Johnson-Shelton, D., Evers, C., & Moreno-Black, G. (2015). Refining the grain: Using resident-based walkability audits to better understand walkable urban form. *Journal of Urbanism, 8*(3), 260–278.

Toh, T. W. (2021, December 19). App helps cyclists plan routes on paths, park connectors. *The Straits Times*.

Willems, T., & Graham, C. (2019). The imagination of Singapore's Smart Nation as digital infrastructure: Rendering (digital) work invisible. *East Asian Science, Technology and Society: An International Journal, 13*(4), 511–536.

CHAPTER 10

Tourism and Mobility

It is customary for tourism to offer promises of escapes from everyday life and enables people to travel to places that are underexplored "often privileged, the exotic and strange reflecting anthropological legacies, to speak of dramatic contrasts between visitors and locals."[1] Increasingly, a growing body of research has combined tourism into the fabric of daily life, including ensuing efforts to de-romanticise tourism. This includes the emerging proximity tourism and slow tourism and by people visiting a city for professional purposes.

Another research strand was to delve deep on more proximate alternatives in tourism that address the issues pertaining to the increase of carbon footprints and local-tourist divides.[2] In contrast to private vehicle ownership and automobility, there has been a heightened awareness and embrace of walking and cycling in public spaces that permit intimate connection with places and its people.[3] Air travel, indeed has been responsible for contributing large amounts of global emissions. Though airways deregulation is to shave off fixed costs (e.g., company and fuel taxes), it seemingly appears the transport mode for the privileged can result all

[1] Franklin and Crang (2001).

[2] Jeuring and Haartsen (2017).

[3] Te Brommelstroet et al. (2017).

Melissa Liow Li Sa and Sam Choon-Yin, *Sustainable Urban Development in Singapore*, Sustainable Development Goals Series,
https://doi.org/10.1007/978-981-99-5451-3_10

sorts of social and spatial divides or unfairness.[4] Therefore, an advanced understanding of non-motorised transport modes and tourism at the local level, can lead to producing more benefits for a wider community.

While all these looks promising, the pressure on mobility networks and urban tourist centres remain to contest the sustainability of tourism. These challenges are primarily due to the venerable weight of the economic benefits of tourism that creates jobs and revenue in situ that are supported by long-haul travel and the mass effects of tourism destinations. Singapore recorded the highest ever number of international visitor arrivals of 19.12 million in 2019 before the arrival of COVID-19 with the travel and tourism sector accounting for 11% of the Gross Domestic Product. Principal markets are primarily Asian and led by China, India, Indonesia, Malaysia and Australia.[5] As a result, transportation needs and consumption patterns that strongly differ from the local social fabric and economic state may undermine tourist cities. An unorthodox means towards the tactical and strategic consequences of tourism is necessary.

Tourism studies have evolved from just being ways to reach a destination. Studies have incorporated a contextual component (transport), and that conceivably semantic to life courses tied with daily commitments, social duties, while offering a distinct collection of mobile experiences.[6] Broadly, changes the perception of the traditional role of tourism linked with *escapades* to *everyday frugality and care*. Policies would require retrofitting since tourism movements are still very entrenched to car and plane travel to get to faraway places which is widely available to the prospective tourists. By tackling the active transport methods and its deduced experiences, as well as multi-modal realities, it is important to link travel means to better fit the local debate surrounding tourism mobilities. Based on the landscape of diversification of tourism and its attractions, ageing seniors and the decisive role of mobility in their welfare, this chapter combines the tourism and leisure aspects in respect to the mundane lives of individuals from various age groups. Exploring tourism activities closer to home enables the integration of

[4] Beaverstock et al. (2016).

[5] World Travel & Tourism Council (2022).

[6] Lumsdon (2000) and Murray (2015).

tourism in active mobility trends. Active mobility such as walking, and cycling are known to specifically suit the density and size of urban nodes and centres.[7]

Walking Tourism

Walking is recognised as being the most basic form of human mobility, and a prevalent practice in tourism and leisure. It has earned more interest in tourist studies as the central activity practised by tourists and an active transport to expedite other activities. Walkability also known to impact visitor experiences in urban contexts. Tourists roam on foot and get to experience unique experiences of city landscapes.

Nonetheless, cities can be complex entities covering a mixture of districts of lesser or greater appeal for tourists and some unfavourable and less preferred for walking. There are also distinctions when one uses a dedicated path informed through mobile applications or in traditional media, and when participating in guided walking trails. Tourists tend to move from one specific destination point to another and roam about the streets at random. Therefore, the size and layout of a city assume a part in comprehending how *walking tourism* can blend with the city's public transport network and links to the tourist accommodation sites. Ganga Samarasekara, Kiyotaka Fukahori and Yoichi Kubota found that the main predictors of tourist walking decisions are potential for activity and exploration, environmental quality, comfort and safety though walking behaviour can be very much personalised for tourists as for residents.[8] Factors such as age, gender, physical fitness, personality, stage of day, weather, duration of stay and the profile of travelling buddies can affect a tourist's ability and preparedness to walk, the selection of attraction places to do so and time undertaken to complete the walking journey. Tourists possibly are less knowledgeable about their surroundings as compared to residents and city workers, the time they have on hand and the frequent need for washrooms and rest areas. An increased risk of accidents may happen to inbound tourists relating in part to victims looking the wrong way and the bus lanes immediately adjacent to walking

[7] Richards and Wilson (2006) and Heinen et al. (2010).

[8] Samarasekara et al. (2011).

pavements.[9] However, not all tourists are unfamiliar to the city in which they are walking and the other influence to walking behaviour would be their purpose of their visit. Urban planning and transport decisions from the governments shape the nature and appeal of the walking environment for tourists..

Indeed, public transport in Singapore has rendered a better helping hand in promoting walking among the tourists to shorten their first to the last-mile (FLM) journeys, for example, from MRT and LRT stations to tourist attractions and place of stay. By 2030, LTA's target is to extend the rail network to 360 kilometres. Singapore will then have a total rail length that is at par with New York City and London, and longer than Hong Kong and Tokyo.[10]

Taking into consideration walking in Singapore's hot and rainy weather can be tough for tourists, Singapore's LTA has implemented several plans to offer a more seamless travel between place of interests and public transport hubs such as bus stops and MRT stations, while continuously upgrading the infrastructure for greater convenience, safety and comfort. Some 200 kilometres of sheltered walkways have been added under the Walk2Ride programme nationwide since 2018. Shelters are provided for high pedestrian traffic spaces. Tourists can engage in sheltered walks within 200 metres from bus interchanges and 400 metres from the MRT stations, LRT stations and designated bus stops with high passenger movement. Other than the sheltered walkways, under the Walk2Ride programme, more lifts have been fitted at over 40 overhead bridges. Overhead bridges and pedestrian crossings have become a permanent structure and pathways are built along most roadways.

Predominant condition in Singapore is the government willingness to invest in amenities and infrastructure that comprise tourism assets. High levels of security and safety that support tourism is a must and every

[9] Baldwin et al. (2008).

[10] LTA is working on some of the major rail expansion projects between 2020 to 2032. The Thomson–East Coast Line (TEL), a medium-capacity MRT line, will be extended in 4 stages (Stage 1 to stage 4), from 2020 to 2024. TEL Stage 5 will be activated in 2025 at Bedok South and Sungai Bedok as well as the Downtown Line (DTL) Hume. In 2026, Circle Line (CCL) will encompass Keppel, Cantonment and Prince Edward Road. The rail extension for Northeast Line (NEL) and the Downtown Line 3 (DTL3) is targeted for completion by 2030. Cross Island Line (CRL) stage 1 and CRL-Punggol Extension are slated to complete by 2030 and 2032, respectively. See Government of Singapore (2021d).

city needs to assure visitors that they can walk on the streets during the day and night without fears of crime. Most general activity concentrates surrounding the tourist attractions and accommodation sites; and walkability happens in free access sites rather than the paid ones. Tourists flock to sites like Little India, Chinatown and the 2.2 kilometres Orchard Road stretch to patronise its hotels, shopping malls and restaurants. With some perfect timing, tourists can get the piece of live action of epochal events such as *Chingay* Parade (an annual street parade held in celebration with the birthdays of the Goddess of Mercy and Chinese deities as part of the Lunar New Year festivities) and Thaipusam (a festival celebrated by Hindu Tamil community) typically in months of January or February and the Great Singapore Sales in June to September. To encourage walking tourism, the Urban Redevelopment Authority (URA) has strived to make walkable and pedestrian friendly by planting tree-lined wide tropical boulevard effect, making its streetscape a vibrant and an attractive place.

URA has set their sights to make the city more pedestrian friendly by linking the above and below ground paths which make a more effortless walk to the closest MRT station with accompanying retail activity. Underground air-conditioned walkways linking to the MRT is appealing for tourists. Notable examples include the underground facility connecting City Hall MRT to Citylink Mall, Raffles City, Suntec City and Marina Square.

Participants exercising outdoor activities such as mountain climbing and kayaking often engage in long walks, consequently adding the broad tourist experience. In terms of tourism mobilities, rhythmanalysis is popularly used for investigating the relationships of walkers, places and their walking experiences. This method analyses the rhythms of urban places and the causes of those rhythms on the occupants of those places. Walking practices and the values are laden by intermingled historic and socio-cultural discourses.[11] Applying rythmanalysis has contributed new research arms that informed placemaking, offering insights of women walking alone on vacation and their experiences coupled with their feelings of vulnerability. Other studies investigate the gendered beliefs of safety fears and skills of women in co-ed hiking communities, as well as individuals with atypical bodies (plump or idiosyncratic physique)

[11] For details, see Küpers and Wee (2018) and Edensor (2010).

and ethnic mobilities.[12] Despite its small size, Singapore offers tourists options to explore adventure and nature. The reclaimed land of Semakau Island was originated slated to meet Singapore's waste disposal needs. At present, the island is used for nature conservation to protect the significant coral reefs in the area, as well as its intertidal habitats and marine life. Mangrove restoration and the construction of silt curtains to minimise sediment damage of nearby sea grass, reefs and sandy shores are also on Semakau Island's agenda. The landfills have been converted into parks, and partially accessible to the public for recreational activities such as intertidal walks, stargazing, birdwatching and fishing.[13]

Cycling Tourism

Cycling tourism offers encouraging impacts such as enabling close connection with nature and interaction with local population even in remote regions, but also contribute to health and welfare. There are more than a few modes of cycling tourism consisting of short haul exploration in rural and urban sites to long-haul package cycling excursions and self-directed touring. These modes are different based on the two-wheeler that the cyclists use which include touring bike, mountain bike and e-bike. One distinctive trait of autonomous cycling tourism is that cyclists prefer their own two-wheeler and their accompanying gear. Cycling tours are often planned alongside cycleways or iconic trails. It is likely for mass tourists to slot an independent cycling tour into their holiday retreats. Therefore, Singapore can be marketed for its cycling tracks for tourist exploration.

In Singapore, tourists who are unfamiliar with bike trails can subscribe to the licenced travel agents to offer good recommendations to *must bike trails*. Other than well-maintained bike wheels and gear, the licensed operators must ensure proper insurance coverage, luggage storage facilities, provide refreshing cold water as well as shower and toilet facilities. *Let's Go* is an award-winning travel agent that has been curating unique travel experiences offering an all-encompassing range of well-designed experiential tours. Tourists can opt to go for its signature cycling tours

[12] Brown et al. (2020), McAnirlin and Maddox (2022), Sarmento (2016) and Stanley (2020).

[13] Connolly and Muzaini (2021).

such as a Historical Singapore Tour, Bike and Bites Food Tour, Singapore in 1920s—Cultural Trails of Tan Ah Huat, an early Singapore immigrant and Marina Bay Night Tour.

The rise of new forms of urban mobility attracts the attention of city planners to provide solutions for the tourist mobility issues while weighing the impacts of public space overcrowding, residents' welfare, and the making of unique tourist experiences.[14] The LTA strives to make Singapore a well-connected cycling friendly city and to its belt, there are 460 kilometres of cycling path. The cycling path network will be expanded to 800 kilometres by 2023 and the cycling network will triple to more than 1300 kilometres by 2030. Some attractions featuring the Lone Tree at the Jurong Lake District, Admiralty Park Playground with the tallest and longest tube slides in Singapore, and Coney Land when riding on #ExploreSGRide cycling routes. These cycling paths will link residents and tourists from homes and tourist spots to bus interchanges, MRT stations and shopping malls. Cycling Path Networks (CPNs) have been fitted with infrastructure improvements to offer a conducive and safe environment for active mobility users from different age groups and competency levels to travel around towns with more convenience and ease.

There are several cycling routes under the #Explore SGRide campaign for tourists to explore. One can scout the mangrove and forest habitats at Admiralty Park in the Woodlands route. Other options include enjoying the island city's skyline along the Marina Bay Rochor route or feel the sea breeze along the Pasir Ris coastline when on the bike. Cycling friendly infrastructure such as improved parking spaces with consideration of convenience and safety aspects are some of LTA's plans beyond the development of the Cycling Path Networks (CPNs).

Other forms such as the cycle rickshaw in some tourist spots are gaining popularity. People ride on rickshaws that are reminiscences of the past. In Singapore, the *Trishaw Uncle Guided Tours* are popular with tourists. Tourists are taken back in time on a nostalgic trip meandering the Singapore streets on a cycle rickshaw, where *Uncle* would share interesting stories of the local and migrant communities who once lived in these

[14] It is a norm for tourists to cycle in popular tourist spots, such as Lisbon, Paris and Vienna. Cyclists navigate the urban city spaces which they have been increasingly dependent on their mobile audio devices. This forms an integral urban cycling culture that are practised by many urban cyclists.

places. Tourists get to witness the remarkable cultural shifts Singapore went through from past to present alongside Chinatown, Little India, Kampong Glam and Singapore River. Tourists can also experience close-up discovery cultural trails and get to drop by shops to get hold of kitschy souvenir.

Slow and Proximity Tourism

The more proximate tourism experiences are seen converging with the daily mobilities, in the temporal and spatial gamut. Underpinned by the mobilities concept, the mobile daily life activities would weave unique leisure and tourism experiences as well as institute community engagement.

There are a few possible reasons.

First, people engaging in the outdoor active mobility can basked themselves amid the natural and urban features that affects one's psychological welfare and enhance his or her quality of life. Emerging research strands of tourism, nature and welfare as a sophisticated nexus begun thriving due to more attention on geographical work, welfare and human and fauna connections of a particular habitat, region and geological period.[15]

Second, tourism studies that examine how tourism is set in people's daily lives, offers a springboard for forthcoming debates on broader topics in regard to people welfare and mobility debates. Examples include tourism estate developers who are inclined to building tourism infrastructure that provides lasting and meaningful benefits to improve the welfare of individuals and destinations rather than offering the less permanent and transient episodic escapades. Mobility envisaged during an activity, encounter and journey or mood while on the move facilitates the pursuit to improving welfare of merely the act of visiting places and the movement itself, severely deviates from just having fleeting hedonic pleasure forms of welfare and tourism.[16]

Third, for economic, social and ethical purposes, the ageing population becomes a more obvious intent of study in mobility research. They represent a more healthy and wealthy population segment. The growing number of people ageing in their usual living environment will impact

[15] Dolnicar et al. (2013) and Fox and Gee (2017).

[16] Cass and Faulconbridge (2017), Creswell (2010) and Smith and Diekmann (2017).

the public spaces and the mobility networks, and to recognise that their mobility may be hampered by ageism resulting from their decline in cognitive and physical functioning and engagement with other people in transport locales.[17]

The initiatives undertaken by the Singapore government as described earlier have effects on the walking experiences of residents of all ages as well as tourists. Tourists are likely to engage with locals in Singapore's underground walkways and pedestrianised streets. For instance, two of four original road lanes in Bencoolen Street are converted into broader cycle track and walkways. There are linkways that are sheltered with natural greenery. This permits mingling of tourists and residents of all ages in Singapore heartlands. Three adjacent streets have been likewise converted. Anderson Bridge known to be the gateway to the arts and cultural district, perching over the Singapore River has been pedestrianised along with another four nearby new pedestrian-friendly spaces. Car-free Sundays in the historic Civic District and Central Business District have been well embraced, stirring a range of events for residents and tourists to enjoy together. Enterprises and community groups play a pivotal role for getting the official project approvals and funding for URA's people and tourism projects.[18] One can look forward to project measures that include increased street activities, traffic lane reductions and additional crossing spaces, and even full pedestrianisation. Some streets have been designated to prioritise pedestrian walking in Singapore's iconic ethnic enclaves. The street market made up of four traffic free streets (Pagoda, Sago, Smith and Trengganu) in Chinatown can be directly accessed from Chinatown MRT station. Certain areas of the street market are air conditioned and equipped with high glass ceilings as part of the refurbishment efforts, partly driven by tourism, but mostly to create a livelier and comfortable public space for anyone.

Proximity tourism deviates from the conventional measures of dissimilarity and distance. The concept pivots on the need for redevelopment of daily spaces as a recreational and respectful environment.[19] Visitors self-identify their own micro-level pursuits which allow them to bask in the community activities they call touristic while gaining the social experiences

[17] Hjorthol (2013), Klimczuk (2015) and Nordbakke and Schwanen (2014).

[18] Kaur (2016).

[19] Diaz-Soria and Llurdes-Coit (2013) and Diaz-Soria (2017).

they desire for. "*Slowing down the pace*" would have room for all ages in such ecosystems and resonates the tenets of slow tourism, while penetrates tourism imaginary. Slow tourism allows to draw upon consideration of residents and the use of environmentally friendly transport modes. This is short of genuflecting any sorts of connecting ageing seniors and slowness. Rather, it works towards pulling slow tourism nearer to the mainstream of contemporary tourism studies or what it is coined as *proximity tourism*.

Proximity tourism has been playing a significant role due to Singapore's historic ethnic enclaves and heartlands.[20] Tourists could interact with locals who have been dwelling, running their family businesses for generations alongside with modern enclaves nestled in Chinatown and Little India and neighbourhoods. Much conservation efforts to preserve the heritage buildings with new greenery sprawl on the high modern hotel and office buildings, makes *walking tourism* more comfortable to partake. *Slow tourism* is suitable for all ages and improves inclusiveness of the very young and aged to participate in walking through these historic ethnic enclaves and places which are mostly unreachable by bus coaches. The Malay quarter of Kampong Glam is one of the most celebrated historic enclaves in Singapore's destination marketing, representing the city-state's multicultural heritage.[21] The five footway is a common place feature ascribed to commands by Sir Stamford Raffles, who declared Singapore as a trading post for the East India Company in 1819, to a town planning committee. The covered colonnaded path is about five feet wide that forms a subset of the shop-house structure and is paved along the front.[22] Tourists and locals are able to stroll and roam in Chinatown and Little India designated conservation spaces and sanctioned buildings. The adaptive use of these sanctioned buildings is carefully preserved so as to retain the original authenticity and façade. *Proximity tourism* blend with *slow tourism* prevails well in the three historic enclaves and Orchard Road often featured as *free and easy* and lots to discover with its 22 themed walking trails promoted via the STB's website and the *Passion Made Possible* branding campaign. Walking is the best way for tourists and locals of all ages to relax, unwind and slow down their pace while getting to genuinely appreciate the natural attractions in this garden city. One of the

[20] For details, see Henderson (2013, 2014, 2018).

[21] Chang and Teo (2001).

[22] Yuen and Chin (1998).

most promoted UNESCO World Heritage sites, is none other than Singapore's very own Botanic Gardens. Other landmarked walking and cycling attractions for tourists are Singapore Riverbanks, coastal spots and man-made theme parks such as the Universal Studios Singapore. Tourists have been increasingly looking for short-term accommodation rentals such as those offered by Airbnb that provide *home away from home* experiences. Back packing has been popular for tourists who stay overnight in boutique hotels and budget hostels in outlying places which possibly need longer walks to these accommodation sites.

STB's Segmentation, Targeting and Positioning Strategies

STB's pursuit of Singapore stories in its travel marketing plan 2016 to 2020 is a form of *soft selling* the destination. One of the purposes is to entice high-yield tourists or visitors to see another side of the country. During the five-year roadmap, STB implemented concrete travel marking plans to cater to changing tourist demographics, including tech-savvy millennials and nature lovers, to offer new tourism market development opportunities. With increased competition for tourists dollars, STB introduced new tourism attractions to overcome the perception of Singapore being a concrete jungle. The STB report "Marketing Strategy: of stories, fans and channels" focuses on the shift towards digital marketing and the importance of storytelling in promoting Singapore as a travel destination, and it underpinned by three strategies—Telling A Great Singapore Story, Targeting the Right Fans and Enhancing our Delivery.[23] The report highlights the need for personalised and targeted marketing strategies that appeal to different segments of tourists. It also emphasises the importance of engaging fans, including influencers on social media platforms, as well as the use of data and analytics to optimise marketing campaigns to build advocacy and consumer brand loyalty). Examples include the can-do-spirit hawker culture cum young entrepreneurs, walking/jogging/cycling monitored by the Healthy 365 national campaign, and the country's obsession of its culinary delights). Overall, the report underscores the need for a holistic and integrated approach to destination marketing that leverages the power of digital channels and compelling storytelling

[23] Singapore Tourist Board (2016).

to attract visitors to Singapore. The STB describes the targeted tourism market segments by needs, income and travel preferences, and applies *psycho-demographic* information to define Singapore's targeted global tourism market segments. This enables STB to formulate motivating and relevant contents for each segment. Four targeted tourism market segments have been identified by the STB. They include (1) Working millennials (age 20–34, employed couples and singles) (2) Families with young kids (adults travelling with kids of 12 years and below of age) (3) Active Silvers (Retirees, 60 and above years of age) and (4) BTMICE or business travellers. *Walking tourism, cycling tourism, slow and proximity tourism* have been found a good fit for the four identified tourism market segments.

Destination Singapore copes with a very tough marketing environment. The working millennials, families with young kids and BTMICE or business travellers are becoming more distracted, technologically-savvy and subsequently more demanding. Disruption is rapidly turning out to be the millennial norm and yet there is a precipitously growing Active Silvers segment, causing in a contrast of worlds and a sudden increase in marketing developments. Singapore experiences strong competitive pressure for the tourism dollar as well as on the marketing front, more destinations are promoting via many channels to gain the multi-screen consumers' attention. A series of marketing developments in Singapore has bolstered the nation's destination attractiveness in recent years. Major developments such as staycations, Singapore's very own cultural and gastronomic walking trails in its ethnic enclaves (Chinatown, Little India and Kampong Glam) and soft adventure by walking and/or cycling in Singapore parks and its beautiful night skyline, can offer tourists a once-in-the-lifetime experience. By leveraging Singapore's current strong suite of products and experiences that embraces *walking tourism, cycling tourism, slow and proximity tourism*, Singapore can continue to attract the right audience particularly the four identified market segments (see Table 10.1).

Applying the geographic segmentation base and the market/product expansion strategies coined by Ansoff's (1957) matrix,[24] STB reported that almost 80% of visitors to Singapore are from Asia Pacific region. This is close to *putting all eggs in one basket*, so there is a need to protect and

[24] Ansoff (1957).

Table 10.1 Tourism market segments

Tourism Market Segments	*Definitions*	*Reasons why a good fit to Singapore's offerings of walking tourism, cycling tourism, slow and proximity tourism*
Working millennials (age 20–34, employed couples and singles)	Millennials or Generation Y are those born between 1981 to 1997, making up for 27% (or 2 billion) of the total global population. In 2020, they form 50% of the global workforce and exposed to online resources, social networks and digital devices[25] These Gen Yers travel an average 35 days per year and are passionate about travel and seek high customer service satisfaction[26]	Millennials travel to broaden their horizons by seeking authentic experience, basking in the local culture, and gaining novel experiences while breaking from their daily routines In this internet-enabled city, Singapore's infrastructure permits the millennials to stay connected and retrieve information at their fingertips. Singapore's local neighbourhoods and vibrant cultural precincts offer exhilarating experiences of different cultures and opportunities to indulge a range of local cuisine. Singapore's entertainment scene and lively nightlife offer lots of avenues to connect with locals, enjoy and meet new people Millennials embrace sustainable travel trends and feel responsible for the surrounding environment. They predominantly support tourism businesses and activities that revolve around sustainable operation models which are turning into "greener" communities. They are likely to indulge in walking tourism, cycling tourism and proximity tourism due to these responsibility callings when travelling within the Singapore island

(continued)

25 VinHMS (2022).

26 StayNTouch (2022).

Table 10.1 (continued)

Tourism Market Segments	*Definitions*	*Reasons why a good fit to Singapore's offerings of walking tourism, cycling tourism, slow and proximity tourism*
		With one out of four leisure visitor arrivals are millennials, the market segment is not only high yielding but would be Singapore's future families with young kids and active silvers in the pipeline[27]
Families with young kids (adults travelling with kids of 12 years and below of age)	Family travel statistics disclose that families incline to travel more than singles or couples who like doing things together and spending quality time with one other. An average family spends US$3,835 annually on travel. 62% on millennial travellers bring along their children of age 5 and below, and it is projected that in the next two years, more family vacations (44%) rather friends (40%) would be travelling together[28]	Family holidays in Singapore is perceived time for bonding with children and create many shared, sweet memories. During staycations and when a family travels overseas, the family members can gain new experiences. Lots of planning are made by the mother since her children's hobbies and interests would be a major influence in the degree of walking and cycling, as well as participating in slow and proximity tourism activities. Young children can get close to the nature and culture during these outside-the-classroom/outdoor educational experiences that can inspire their curiosity of the world.

[27] Singapore Tourism Board (2016).

[28] Williams (2022, December 9).

Tourism Market Segments	*Definitions*	*Reasons why a good fit to Singapore's offerings of walking tourism, cycling tourism, slow and proximity tourism*
		Singapore is well positioned to attract family travellers. STB gathered 66% of its visitors from the top 14 geographical markets found Singapore is one of the popular family destinations in Asia. The growing family market share is credited to the travelling ease within the destination, the nation's multicultural heritage and rich edutainment experiences. There is always something for everyone, making the Singapore Destination very appealing to families.[29]
Active Silvers (Retirees, 60 and above years of age)	Surveys conducted by the Silver Travel Advisors gathered that 64% of active silvers are females and 36% are males. The active silvers normally plan for travel in the next three months. 74% of these active silvers tend to take more than one holiday in a year and 72% prefer to go for shorter breaks. 22% have cited spending more now than the pre-pandemic times. They are going for more trips to faraway places (30%), once-in-a-lifetime trip (24%) and multigenerational holidays (16%). 56% consider sustainability and responsible travel when making travel decisions. Example, travel using various transportation means that generate less carbon footprint (carbon-neutral), that does not affect the local people or environment in a detrimental way[30]	Higher spending is recorded from this Singapore's tourism market segment that comprised 16% of the leisure visitor arrivals in 2014. They are mostly from Germany, United Kingdom and United States of America.[31] The walking tourism, cycling tourism, slow and proximity tourism would be appealing to these active silvers who prefer a slower pace. With this comfortable pace, they get to indulge in a wide range of food and the local culture in Singapore ethnic enclaves (Chinatown, Little India and Kampong Glam). Active silvers who prefer soft adventure in Singapore's nature can do walking and/or cycling in its parks (the once-in-the-lifetime experience)

(continued)

29 Singapore Tourism Board (2016).

30 Silver Travel Advisor (2022).

31 Singapore Tourism Board (2016).

Table 10.1 (continued)

Tourism Market Segments	*Definitions*	*Reasons why a good fit to Singapore's offerings of walking tourism, cycling tourism, slow and proximity tourism*
BTMICE or business travellers	Business travellers travel to various locations for attending business meetings, conferences, exhibitions, trade shows; investment purposes; meeting business partners or suppliers, opening new company branches or offices and many more A business traveller travels on average 9 times in a year. Average length of stay is between 6 and 28 days. The younger workforce perceived as one of their job perks, and even as a status symbol (65%) to be able to travel for work There is an emerging trend of bleisure travel (Business + Leisure) especially among the younger workforce who are looking to do more than work during their business trips. Business travellers are more open to stay in unconventional accommodation options (such as home-like apartments and small boutique hotels) rather than the conventional hotel chains, away from the city centre, as slow pace and proximity to leisure activities take over in forms of priorities during their trip planning in the later part of the day. The younger business travellers, especially the millennials, prefer to do self-booking and go outside employer-approved channels when booking transportation and properties for their trips. Known as *flipping the trip script* and changing corporate travel policies have seen bleisure trips getting more unconventional, extended, and boosting in-person connection[32]	Singapore has been the top choice for business travels due to its sophisticated and complete business infrastructure. The Walk, Ride (on Mass Rapid Transit or MRT trains) and Cycle programme allowed the busy business travellers to get where they need to go conveniently. This form of bleisure offers business travellers lots of opportunities to experience tourism products beyond their daily business engagements Be it for the Singapore's beautiful night skyline, its rich cultural and gastronomic delights, built heritage (e.g., museums, religious temples, mosques, colonial buildings) and natural heritage attractions such as its parks, marshes, wetlands and islands are just a few stones away. Business travellers get to unwind and embrace slow and proximity tourism activities after the whole day at work

32 Airssist (2022).

diversify the tourism market demographics to sustain Singapore's tourism industry growth. Other than strengthening the *market penetration* in the conventional source markets, mainly the Tier 1 cities in Australia, China, India, Indonesia, Japan and Malaysia; *market extension* efforts are intensified in the strong visitor arrival growth markets in Tier 2 cities of China, India and Indonesia markets. STB is also looking at expanding Singapore's visitor base and diversifying the market portfolio in other regions. STB has already start embarking pipeline projects to increase *market penetration* in long-haul markets such as France, Germany, United Kingdom and United States.[33]

Tourism and Sustainable Development Goals

Tourism is an instrument for change. For families, social groups and individuals, tourism is a way to enjoy a change of routine, take a break from everyday life and connect with new people and places. Tourism are channels for making a living for businesses and entrepreneurs. Governments and authorities perceive the tourism industry as a tool for new economic activity creation and generate jobs, leading to regional development for tourist service-based nations such as Singapore. Many policymakers have classified the tourism industry as important levers for enabling sustainable development into practice.

The connection between the tourism industry and sustainable development has given rise to sustainable tourism, which takes into account the impact of tourism on the environment, while striving to provide a positive experience for visitors. It involves promoting responsible tourism practices that preserve natural resources and protect local heritage. As tourism is primarily driven by the need to serve visitors and tourists, it can have both positive and negative impacts on the environment. Achieving a balance is crucial. To balance the positive and negative externalities for environment and Singapore's localities, *walking tourism, cycling tourism, slow and proximity tourism* have been integrated in its Singapore's Green Plan 2030, and in support of the United Nations Sustainable Development Goals (SDGs), particularly (1) SDG 8—promoting inclusive and sustainable economic growth, decent work, full and productive employment (2) SDG 12—maintaining sustainable production and consumption tourism

[33] Singapore Tourism Board (2016).

trends (3) SDG 14—sustainable use and conserve pristine condition of seas, oceans and marine resources.[34]

Singapore has long been known for its urban cleanliness, safety and efficiency, but it now also positioning itself as a walkable city that offers a variety of pedestrian-friendly attractions and experiences. The STB has identified walking tourism, cycling tourism and slow and proximity tourism as key genres to appeal to high-yield tourists and align with Singapore's vision of becoming a "City in a Garden". Singapore understands that sustainable development requires more than industry self-regulation. To avoid tourism taking place for the benefits of tourists and not locals, strict regulatory measures, through the STB and other agencies are put in place to prevent overconsumption of resources in various destinations across the island.[35] Singapore has from the early beginning, positioned itself within the framework of international politics, capital and technology. As it seeks to embrace the economy as a global city, a talent hub and a centre for innovation and enterprise it will increasingly have to play even closer by the rules of a global marketplace. The Walk, Ride and Cycle and green infrastructure are gradually expanded to cope with an increasing number of tourists and local visitors in Singapore, but it may not truly fulfil the needs of sustainable development, i.e., the well-being of people, quality of life and the environment. Let's take Singapore's Integrated Resorts as an example. At the first instance, the construction of the Integrated Resorts seems beneficial to the Singaporean society. Both local and foreign interest has increased with the manifold recreation and entertainment benefits that the integrated resorts can offer. It creates lots of opportunities for people to engage in the four forms of tourism genres (*walking tourism, cycling tourism, slow and proximity tourism*) within the urban green infrastructure of the integrated resorts. Singaporeans also get to benefit from some growth of its local economy that reduces the government debt and better public services. We should also ask the

[34] From the sustainable economic growth viewpoint, tourism helps to increase international investments and trade, Gross Domestic Product growth and drives Walk, Cycle and Ride (WCR) infrastructure to support a sustainable tourism environment for Singapore. However, it can be challenging as it appears that the correlation between development and growth ideology, often linked with the neoliberal economic thinking. Potentially, it can be in conflict with other SDGs, such as decent work and economic growth (SDG 8), reducing inequality (SDG 10) and undertake climate action (SDG 13).

[35] Tourism is a growing worldwide industry, "its unsustainability behind a mask that is all the more beguiling because it appears so sustainable" (Hollenhorst et al., 2014).

types of jobs generated relating to SDG 8 (Decent Work and Economic Growth) and SDG 10 (Reduced Inequality). According to figures from Channel News Asia in 2019, the two Integrated Resorts employs 20,000 people of which more than 65% are locals—which is good. Nevertheless, by legalising gambling had also bagged social ills such as gambling addiction, illegal money lending by loan sharks[36] and money laundering by scammers.[37] To encourage more responsible gaming entertainment among locals, an increase of casino entry levies was imposed on locals in April 2019. Such regulatory controls had been effective as the number of visits of Singaporeans and Permanent Residents to the casinos had since dropped substantially.[38]

The construction of the Integrated Resorts had transformed Singapore's skyline but also serve as a part of repositioning Singapore's brand and uniqueness.[39] Radical changes to the natural environment due to the Integrated Resorts construction have been documented. Such massive development that nestled right in the heart of the city, however, undertook embankments and land reclamation along the Marina Bay in the 1990s. Concrete ways were instituted to mitigate the threats to the physical environment by enabling the four forms of tourism genres in the adjoining Gardens by the Bay which supports Singapore's ambition to be a global city. The Marina Bay development completed in 2006 and was helmed by the following vision by then Minister for National Development, Mah Bow Tan who cited in 2005 that "it is to be a place for thriving business, gracious living and endless entertainment, as well as a

[36] 113 people were under investigation for loan shark activities after week-long blitz by the police between the period of 23–30 May 2022. See Ng (2022, May 31).

[37] A 52.9% increase in scam cases drove up the total number of reported crimes to 46,196 in 2021. 23,931 scams worth S$504 million (US$353 million) were reported in 2021. A total 14,349 scams worth S$227.8 million (US$160 million) were reported for the first half of 2022 (Cusack, 2022, September 25).

[38] A 50% increase in the 24-hour casino entry levies from SG$100 to SG$150 and annual levy from SG$2,000 to SG$3,000 were imposed by the Casino Regulatory Authority (CRA) in April 2019, as part of the expansion agreements signed with operators Genting Singapore and Las Vegas Sands. The change had resulted in a significant decline in the number of visits made by Singapore Citizens and Permanent Residents, down from 4.0% in FY 2018 to 2.7% in FY 2019 as reported by CRA Chairman Tan Tee How in the regulator's 2019/2020 Annual Report (Newsdesk, Inside Asian Gaming, 2020, November 2).

[39] Kosasih (2019).

place of creativity, arts and culture, architectural excellence, world class recreation, global commerce and modern city life rolled into one". To reduce the climate change effects (SDG 13—Climate Action) and sustain SDG 14 (Life below Water), the Marina Barrage was developed that comprise of the Kallang Basin, Singapore River and water body within Marina Bay to setup a controlled water basin. For tourists or local residents who are keen to enjoy the beautiful green infrastructure of this waterfront city in a garden, can take a 12 kilometre walk around the whole bay area while engaging *in walking tourism, cycling tourism, slow and proximity tourism.*[40] The Marina Barrage has a common services tunnel that acts as a primary conduit that provides ample space for the pipes supplying water, underground electrical cables and district cooling systems (as in contrast to individual building air conditioning). These environmentally friendly design solutions allow further fulfilment of SDG 6 (Clean Water and Sanitation) and Affordable and Clean Energy (SDG 7) as well as for sustaining proper maintenance for its 54-ha central park.

Despite the concerns on climate action (affecting SDG 13), SDG 6 (Clean Water and Sanitation) as well as Affordable and Clean Energy (SDG 7) due to the iconic Marina Bay development that locates the two integrated resorts, Singapore city has come a long way. With judicious adaptation of environment-friendly solutions, it has posed an exemplary model for other cities on how sustainable tourism and related SDGs can be achieved together. It is evident that a strong political will to institute sustainability targets is pivotal to lead to success. On this point, Singapore's has the potential to present itself as a textbook example of a good sustainable city environment.

Final Thoughts

Tourism is expressed to many, as a reprieve from the hustle and bustle lifestyle in these modern days and located *off the beaten* tracks at locations not normally visited. Turbocharged by pressures on mobility networks and urban tourist centres, there is an increasingly interest to explore the likelihood of tourism closer to home. The investigation of tourist experiences is often influenced by the motorised public transport. This chapter

[40] Wong (2008).

combines two strands of literature, starting with defining tourists' mobilities in the forms of walking and cycling. Another contribution is the age-friendliness approach to tourism and the more proximate forms of tourism *within a stone's throw*, all within Singapore's urban landscape. Singapore's LTA played an integral role to plan and execute the sharing infrastructure to enable all-inclusive transport mobility for both residents and tourists in Singapore.

Despite there have been advances in tourist mobility in city destinations, there is still much research foraging that needs to be done. The chase to capturing insights on walking and cycling as ways of active transport can expand our understanding into tourists' mobility inclinations and experiences in the urban landscape. More publications in the non-Western culture are required since extant research on walking and cycling tourism is mostly anchored in the Western tourism and leisure context. Disparate perceptions of repeatedly disregarded, hushed and side-lined walkers and cyclists, involving atypical bodies (plump or idiosyncratic physique) individuals and ethnic mobilities is one research avenue that needs more research attention. Comprehending the experiences and habits of the less likely walkers and cyclists can make researchers be more aware of the barriers to partaking the exercise. Also rendering new routes to unravelling places encountered, made and re-made, amid the broader theories of tourism mobilities. Lastly, research should consider the heightened digitisation of tourism mobilities. Tourists are increasingly sharing their experiences through visuals and audio records while travelling, creating visual *pengalaman* (experiences in Malay) in their own's right and record it. These live(life) streaming recordings reveal features about our food habits, our close and distant others and places that many people visit and enjoy.

While opportunities for tourists to partake in events such as Chingay and Great Singapore Sales can be accessible by walking and/or cycling, there are more things that the authorities can do to enhance *walking tourism* and the tourist experiences. Singapore, like any other country possesses inherent challenges as a city made for walking and cycling. Singapore is the third most densely populated cities hosting a high number of inbound tourists. Even the widened Orchard Road pathways can be overcrowded at times. Little India and Chinatown not only attract tourists but are favourite weekend spots for migrant workers to catch-up with one another. This is when walking and cycling can be hindered due to the massive crowd. Better road crossings, clear signages at Paterson and Orchard Roads, as well as the overhead and underground corridors are important to ensure seamless Walk, Cycle and Ride travel for unfamiliar tourists to reach the intended destinations hassle free. Without clear signages, tourists could easily lose their way when trying to navigate among the subterranean passageways of a shopping mall basement, which would be easier if this is done through a street crossing above ground. Freedom to walk and cycle can be reduced through the continual works to burrow tunnels for new MRT stations, and temporary routes can impede the locals' and tourists' experience when partaking walking and cycling as they would likely be redirected to use temporary paths. Meticulous care to construct proper drainage, anti-slip and even-surfaced floor grating needs to be carried out as it concerns pedestrians' safety especially during and after heavy downpour. Regulations have been enacted to improve pedestrian safety. These are important considerations as means to retain Singapore as a tourism hub, bearing in mind the possibility of tourists sharing their experiences over the social media and travel advisories forums.

References

airssist. (2022). *Top business travel trends 2023*. https://airssist.com/business-travel-trends/. Accessed on 26 December 2022.

Ansoff, H. I. (1957). Strategies for diversification. *Harvard Business Review, 35*(5), 113–124.

Baldwin, A., Harris, T., & Davies, G. (2008). Look right! A retrospective study of pedestrian accidents involving overseas visitors to London. *Emergency Medicine Journal, 25*, 843–846.

Beaverstock, J. V., Derudder, B., Faulconbridge, J., & Witlox, F. (2016). *International business travel in the global economy*. Routledge.

Brown, L. D., de Coteau, D., & Lavrushkina, N. (2020). Taking a walk: The female tourist experience. *Tourist Studies, 20*(3), 354–370.

Cass, N., & Faulconbridge, J. (2017). Satisfying everyday mobility. *Mobilities, 12*(1), 97–115. https://doi.org/10.1080/17450101.2015.1096083

Chang, T. C., & Teo, P. (2001). From rhetoric to reality: Cultural heritage and tourism in Singapore. In L. Low & D. Johnstone (Eds.), *Singapore Inc: Public policy options in the third millenium* (pp. 273–303). Asia Pacific.

Connolly, C., & Muzaini, H. (2021). Urbanizing Islands: A critical history of Singapore's offshore Islands. *Environment and Planning E: Nature and Space*. https://doi.org/10.1177/25148486211051082

Creswell, T. (2010). Towards a politics of mobility. *Environment and Planning D: Society and Space, 28*(1), 17–31. https://doi.org/10.1068/d11407

Cusack, J. (2022, September 25). *Singapore Financial Crime Dashboard 2022 by FCN*. https://thefinancialcrimenews.com/singapore-financial-crime-dashboard-by-fcn-2022/. Accessed on 28 January 2023.

Diaz-Soria, I. (2017). Being a tourist as a chosen experience in a proximity destination. *Tourism Geographies, 19*(1), 96–117.

Dıaz-Soria, I., & Llurdes-Coit, J. C. (2013). Reflections on proximity tourism as a strategy for local development. *Cuadernos de Turismo, 32*, 65–88.

Dolnicar, S., Lazarevski, K., & Yanamandram, V. (2013). Quality of life and tourism: A conceptual framework and novel segmentation base. *Journal of Business Research, 66*(6), 724–729. https://doi.org/10.1016/j.jbusres.2011.09.010

Edensor, T. (2010). Walking in rhythms: Place, regulation, style and the flow of experience. *Visual Studies, 25*(1), 69–79.

Fox, R., & Gee, N. R. (2017). Great expectations: Changing social, spatial and emotional understandings of the companion animal–human relationship. *Social & Cultural Geography, 20*(1), 43–63. https://doi.org/10.1080/14649365.2017.1347954

Franklin, A., & Crang, M. (2001). The trouble with tourism and travel theory. *Tourist Studies, 1*(1), 5–22. https://doi.org/10.1177/146879760100100101

Government of Singapore. (2021a). *Cycling*. https://www.lta.gov.sg/content/ltagov/en/getting_around/active_mobility/walking_cycling_infrastructure/cycling.html. Accessed on 19 September 2021.

Government of Singapore. (2021b). *Joo Chiat*. https://eresources.nlb.gov.sg/infopedia/articles/SIP_946__2008-11-13.html. Accessed on 4 September 2021.

Government of Singapore. (2021c). *LTA annual report 2019/20: Riding it out together*. https://www.lta.gov.sg/content/dam/ltagov/who_we_are/statistics_and_publications/report/pdf/LTA_AR1920.pdf. Accessed on 12 September 2021.

Government of Singapore. (2021d). *Rail network*. https://www.lta.gov.sg/content/ltagov/en/upcoming projects.html. Accessed on 19 September 2021

Government of Singapore. (2021e, July 1). *Does the government have a population target, e.g. 10 million?* https://www.gov.sg/article/does-the-government-have-a-population-target#:~:text=At%20that%20update%2C%20the%20Government,below%206.9%20million%20by%202030. Accessed on 3 February 2022.

Heinen, E., Van Wee, B., & Maat, K. (2010). Commuting by bicycle: An overview of the literature. *Transport Reviews, 30*(1), 59–96.

Henderson, J. C. (2013). Singapore's Little India: Ethnic districts as tourist attractions. In B. Garrod & A. Fyall (Eds.), *Contemporary cases in heritage* (pp. 34–50). Goodfellow.

Henderson, J. C. (2014). Bidding for world heritage: The case of Singapore's botanic gardens. *Tourism, Culture and Communication, 14*(2), 63–76.

Henderson, J. (2018). Making cities more walkable for tourists: A view from Singapore's streets. *International Journal of Tourism Cities, 4*(3), 285–297.

Hjorthol, R. (2013). Transport resources, mobility and unmet transport needs in old age. *Ageing and Society, 33*(7), 1190–1211. https://doi.org/10.1017/S0144686X12000517

Hollenhorst, S. J., Houge-MacKenzie, S., & Ostergren, D. M. (2014). The trouble with tourism. *Tourism Recreation Research, 39*(3), 305–319.

Jeuring, J., & Haartsen, T. (2017). The challenge of proximity: The (un)attractiveness of near-home tourism destinations. *Tourism Geographies, 19*(1), 118–141. https://doi.org/10.1080/14616688.2016.1175024

Kaur, G. (2016, May 20). Street life thrives in car-free zones. *The Straits Times*. www.straitstimes.com/lifestyle/street-life-thrives-in-car-free-zones. Accessed on 25 October 2017.

Klimczuk, A. (2015). *Economic foundations for creative ageing policy*. Palgrave Macmillan.

Kosasih, J. (2019, September 20). *GP extra credit: Ten years after their construction, have the integrated resorts had a significant impact on Singapore society?* https://eunoiapress.wordpress.com/2019/09/20/gp-extra-credit-ten-years-after-their-construction-have-the-integrated-resorts-had-a-significant-impact-on-singapore-society/. Accessed on 27 January 2023.

Küpers, W., & Wee, D. (2018). Tourist cities as embodied places of learning: Walking in the "feelds" of Shanghai and Lisbon. *International Journal of Tourism Cities, 4*(3), 376–390.

Lumsdon, L. (2000). Transport and tourism: Cycle tourism—A model for sustainable development? *Journal of Sustainable Tourism, 8*(5), 361–377. https://doi.org/10.1080/09669580008667373

McAnirlin, O., & Maddox, C. B. (2022). 'We have to be a little more realistic': Women's outdoor recreation experiences in a community hiking group. *Annals of Leisure Research, 25*(3), 335–351.

Murray, L. (2015). Age-friendly mobilities: A transdisciplinary and intergenerational perspective. *Journal of Transport & Health, 2*(2), 302–307. https://doi.org/10.1016/j.jth.2015.02.004

Newsdesk, Inside Asian Geming. (2020, November 2). *Singapore regulator says casino entry levy increase has kept locals away*. https://www.asgam.com/index.php/2020/11/02/singapore-regulator-says-casino-entry-levy-increase-has-kept-singapore-locals-away/. Accessed on 28 January 2023.

Ng, K. G. (2022, May 31). 113 people under investigation for loan shark activities after week-long blitz. *The Straits Times*.

Nordbakke, S., & Schwanen, T. (2014). Well-being and mobility: A theoretical framework and literature review focusing on older people. *Mobilities, 9*(1), 104–129. https://doi.org/10.1080/17450101.2013.784542

Richards, G., & Wilson, J. (2006). Developing creativity in tourist experiences: A solution to the serial reproduction of culture? *Tourism Management, 27*(6), 1209–1223. https://doi.org/10.1016/j.tourman.2005.06.002

Samarasekara, G. N., Fukahori, K., & Kubota, Y. (2011). Environmental correlates that provide walkability cues for tourists: An analysis based on walking decision narrations. *Environment and Behaviour, 43*(4), 501–524.

Sarmento, J. (2016). Tourists' walking rhythms: 'Doing' the Tunis Medina, Tunisia. *Social &Cultural Geography, 18*(3), 295–314.

Silver Travel Advisor. (2022). *Silver travel advisor industry snapshot April 2022.* https://en.calameo.com/silvertraveladvisor/read/0044274231677a12c8b3f. Accessed on 26 December 2022.

Singapore Tourism Board. (2016). *Marketing strategy: Of stories, fans, and channels.* https://www.stb.gov.sg/content/dam/stb/documents/assistancelicensing/marketing-innovation/Marketing_Strategy-Of_Stories_Fans_and_Channels.pdf. Accessed on 26 December 2022.

Smith, M. K., & Diekmann, A. (2017). Tourism and wellbeing. *Annals of Tourism Research, 66*, 1–13. https://doi.org/10.1016/j.annals.2017.05.006

Stanley, P. (2020). Unlikely hikers? Activism, Instagram, and the queer mobilities of fat hikers, women hiking alone, and hikers of colour. *Mobilities, 15*(2), 241–256.

StayNTouch. (2022). *2022 hotelier technology sentiment report*. https://www.stayntouch.com/2022-hotelier-technology-sentiment-report-2/. Accessed on 8 April 2023.

Te Brommelstroet, M., Nikolaeva, A., Glaser, M., Skou Nicolaisen, M., & Chan, C. (2017). Travelling together alone and alone together: Mobility and potential exposure to diversity. *Applied Mobilities, 2*(1), 1–15. https://doi.org/10.1080/23800127.2017.1283122

VinHMS. (2022). *Millennials travel trends redefine the hospitality in 2022.* https://www.cihms.com/en/millennials-travel-trends-redefine-the-hospitality-in-2022/#:~:text=Experiential%20travel&text=An%20Expedia%20survey%20has%20found,at%20premium%20hotels%20and%20resorts. Accessed on 26 December 2022.

Williams, B. (December 9, 2022). *80+ Family travel & tourism statistics (2022).* https://www.dreambigtravelfarblog.com/blog/family-travel-tourism-statistics. Accessed on 26 December 2022.

Wong, T. C. (2008). Integrated resort in the central business district of Singapore: The land use planning and sustainability issues. *Spatial planning for*

a sustainable Singapore (pp. 59–78). https://doi.org/10.1007/978-1-4020-6542-2
World Travel & Tourism Council. (2022). *Economic impact report*. https://wttc.org/research/economic-impact. Accessed on 8 April 2023.
Yuen, B., & Chin, H. C. (1998). Pedestrian streets in Singapore. *Transportation, 25*(3), 225–242.

CHAPTER 11

Modern Architectural City for an Ageing Population

An Urban Systems Study has been in place since the 1980s, encompassing specific domains of the public sector which are responsible for piecing together the policies to transform Singapore into a liveable and sustainable city. The work, housed in The Centre for Liveable Cities (CLC) investigated the intricate trials in chasing the goal of an engaged and a happy senior population who goes on to contribute to the Singapore society and beyond.[1] The government boarded on a year-long consultation process in 2014 comprising some 4,000 residents from different walks of life, to gauge their aspirations for ageing. Guided policymaking and the dialogues around its senior population have become the *talk of the town*. The key is to design strong foundations while the nation gears towards planning and implementing initiatives for its senior population to age well. Through employment and retirement reserves, the seniors gained financial security. The island city-state reformats its physical infrastructure throughout residential and accessibility territories, to enable seniors integrate and be a part of a more resilient and cohesive community.

This chapter highlights the initiatives to improve the physical infrastructures to support the ageing population, and the urban study gaps that

[1] The Centre for Liveable Cities (CLC) was established in June 2008 by the Ministry of National Development (MND).

Melissa Liow Li Sa and Sam Choon-Yin, *Sustainable Urban Development in Singapore*, Sustainable Development Goals Series,
https://doi.org/10.1007/978-981-99-5451-3_11

need to be solved to improve the nation's ageing landscape. The work towards planning to age well has not yet complete and continual partnership efforts from the public, private and community realms is required to sustain the goals to have Singapore's seniors to age well. The altruistic goal is making Singapore future-ready and a *city liveable for all ages* but also to fulfil the lifestyle of its still strong-going senior workforce in this knowledge economy era. The chapter will demonstrate how this approach aligns with Sustainable Development Goals (SDGs) such as Good Health and Well-Being (SDG 3), Affordable and Clean Energy (SDG 7), Sustainable Cities and Communities (SDG 11) and Partnerships for the Goals (SDG 17).

Singapore's Demographic and Social Trends

Singapore has one of the most rapid ageing populations in the world. Singapore's seniors account for 16.0% of the population in 2021, and the median age of Singapore's resident population is 41.8 years in 2021, up from 24.4 years in 1990 and 34.0 years in 2000.[2] Singapore holds the fifth spot in terms of the highest life expectancy in the world at 83.9 years in 2020 from 81.7 years in 2010.[3] The number of working-age citizens for each senior citizen will fall to 2.7 by 2030 from 5.9 in 2013, and by 2030, the seniors will triple to more than 900,000.[4] Henceforth, Singapore is experiencing a shrinking workforce coupled with its ageing society. Other broad trends are getting more visible. The elderly is more independent, well-educated and aspirational. The nuclearisation of families has increased with the number of empty nesters and single elderly to reach 288,000, of which 83,000 will be living alone by 2030. In 1970, 1 in 31 Singaporeans was aged 65 or older, in 2015, it was 1 in 8 and by 2030, it will be 1 in 4 seniors.[5] Seniors above the age of 60 years old with dementia in Singapore are expected to rise over 130,000 by 2030. The prevalence of dementia was found to be 10% in the seniors aged 60 years and above in a nationwide Well-being of the Singapore Elderly

[2] Singapore Department of Statistics (2021).

[3] Worldometer (n.d.).

[4] Singapore Department of Statistics (2019).

[5] Ministerial Committee on Ageing (n.d., -a, -b).

(WiSE) study.[6] These demographic trends will cause considerable forces on the society, not merely on supporting the seniors to lead happy and meaningful lives but would also intensify the needs to offer healthcare. Changing our mindset on how to better manage the nation's physical and social spaces while providing medical and care facilities would be necessary. Other implications would be the impacts on the larger society in terms of daily living amid an older demography.

Developing Singapore *City for All Ages*

Cities began to perceive ageing more as a social rather than a medical issue by mid-2000s. While designing the urban environments, accessibility, mobility and the safety of the seniors must be kept in mind while creating more liveable spaces for the rest of the population. *City for All Ages* is a theme that captures above the standalone facility for seniors. Accessibility to eldercare and community programmes to increase the overall liveability of the built environment are other key considerations. One of the main pillars in mounting the ageing issues in Singapore is known as *ageing-in-place*. It is a vision that foresees the elderly growing old in familiar spaces such as in their home environment and community spaces, set to cause little disruption or change to one's lifestyle, even as their cognitive and physical abilities deteriorate over time. This requires consideration that the living environment would have to be accessible that fulfil the needs of people when they age. It pivots on a conviction that these needs are better provided inside the community that they have been accustomed, rather than growing old in separate and distinct spaces with seniors like themselves.[7] Reimagining the elderly to be able to carry on engaging and meaningful lives is where building a *City for All Ages* thrives on. By implementing age-friendly environment, increasing employability and lifelong learning choices, attention needs to be placed towards enabling *ageing-in-place* irrespective of the individual's health or income status.

Retrofitting the physical environment is essential but not enough to building an age-inclusive city. The *City for All Ages* initiative needs to poise on service integration and programming, just not through hardware. The Singapore government's role has grown from one that provides

[6] Agency for Integrated Care Pte Ltd. (2022).

[7] Chong et al. (2015).

aged care infrastructure as well as doctrines policy, regulatory and financial frameworks to supporting and partnering families and community sector to offer and operationalise services for the seniors. Changes in family structures over time has led to a paradigm move impacting caregiving plans. Dual income source families are on the rise and more females enter the labour force. Women can only spend a small portion of their time they previously give for caregiving, and when a husband suffers a debilitating sickness, the women would take over the role as the sole breadwinner and the labour force needs are shifted to the wives.[8] There are fewer adult children looking after their elders, some of them work and live abroad.

The dwindling size of nuclear families has intensified the need for the elderly to depend on kith other than kin for social assistance. There is an increasing call to shape the kith profile that can include the elderly's peers and form social systems to strengthen inter-generational relationships in neighbourhoods, at work, or within recreational spaces. The Government has undertaken measures to drive more contribution from the private sector to provide senior housing. Market demand led by the seniors would somewhat become a reality. There is a challenge as Singapore experiences high property prices due to its land scarcity. Attractive government incentives to drive these developers to provide housing for the seniors may help to persuade them. However, the experience in 2012 with *Jalan* (street in Malay) Jurong Kechil, the initial plans to develop a retirement village ended up as a conventional housing site in the end. It demonstrates a dearth of understanding and keenness among the real estate developers to provide elderly housing models. A post-mortem exercise is necessary to identify the inhibiting forces. There are several possible reasons such as (1) lagging market reading and consumer demands for private retirement villages or inadequately designed (2) for the operationalisation of the retirement villages and elderly facilities could it be better to be government-led instead of the private real estate developers and (3) government to look in designating a land allocated or apply tender bid that would be the basis for elderly living options.[9]

Survey results show that seniors who are living alone tend to exhibit depression. To support this group of seniors, GoodLife!, an activity

[8] Cheng et al. (2019, September).

[9] Kua (2021).

centre, was setup by the Marina Parade Family Service Centre and have since recruited 40 volunteers that form a befriending network that comprise of other seniors between 53 and 84 years old. Agency for Integrated Care (AIC) was established where it has a community outreach team to support the seniors suffering from depression and dementia.[10] A human-centric and a ground-up tactic is applied across people and public segments, community and government as well as agency lines to deal with the needs of the seniors.

Singapore deals the irrefutable likelihood of a quarter of Singaporeans by 2030, would be aged 65 and above. While the discourse is shifting, the move has been slow. Therefore, there are government considerations to take lead, intervene and operationalise services for the masses while being mindful of the ageing needs to make it happen soon. In this regard, the government has pursued three basic tenets. Firstly, the *Many Helping Hands* concept was coined, where the family will form the first line of help for the elderly. Secondly, the government has taken steps to support by partnering volunteer welfare bodies and social services to deliver on the ground while skilfully implementing the policy framework for ageing and preparing the needed infrastructure. Singapore residents ranked family (73%) in order of their role importance in providing eldercare compared to government, employers, community. This contrasts with Singapore's governance approach that ranked the government itself as responsible in the first or second spot, while placing the community, then family in bearing the duties of care.[11] There seems to be a decoupling of expectations of the Singapore residents and the policymakers on who should care for their elderly. The higher ranking of the government rather than the community shows a potential scepticism in the capability to provide sufficient and proficient support and services for the elderly.

[10] D. Yeo (2013, January 7).

[11] Gee et al. (2018).

Walkability Enabling Seniors

A new phenomenon in the cities around the world has seen an unprecedented surge in car use per capita across developed cities and emerging cities such as Singapore.[12] Cities have looked at ways to decouple greenhouse emissions from wealth generation from fossil fuels and in the scope of urban planning to reduce car usage. Singapore has taken the lead in reducing the number of vehicles on its roads. This effort has proven beneficial for Singapore's ageing population, which accounts for 16.0% of the total population as of 2021. Singapore's elderly can now travel more seamlessly thanks to a well-designed public transport system, as well as walkways and cycling infrastructure which compensates for the reduction in private car ownership.

Singapore is actively working towards revitalising its central business districts and residential neighbourhoods by reintroducing well-designed street patterns that prioritise walkability. Interestingly, even seniors in Singapore, including those with knowledge-based occupations, are actively reducing their reliance on cars, despite having the financial means to own them. In fact, seniors in Singapore are just as receptive as younger individuals to embracing eco-friendly transportation options such as walking, cycling and electric vehicles. This cultural shift is driven by the numerous benefits associated with these modes of transportation, including their affordability, the opportunity for recreational activities and their positive impact on the environment. By emitting fewer greenhouse gases, creating less air and noise pollution and promoting a sustainable lifestyle, these transportation choices align with Singapore's commitment to urban renewal and its endeavour to detach car usage per capita from economic growth indicators. To encourage more people including seniors to ride on electric public buses and consume electric car sharing services, Singapore's urban planning offers sheltered walkways/bus stops, dedicated cycling paths and fast corridor service in its walking precincts with EV recharge stations. Singapore seeks to create a greener and more sustainable land transport sector, cutting down peak land transport emissions by 80% by 2050. The Singapore Green Plan 2030 advocates a strong drive to electrify Singapore's vehicle population, which would help the nation to achieve its vision of 100% cleaner energy vehicles by 2040.[13]

[12] Newman and Kenworthy (2015), Koh and Lee (1994).

[13] Government of Singapore (2022, April 8).

The steady increase in bicycle use with 4.0% of its population includes Singapore seniors who enjoy its recreational benefits together with peers and the younger generation. Middle-aged professionals who work in the Central Business District or who live in the eastern coastal districts, such as Siglap or Marine Parade are enticed to the lifestyle perception of commuting autonomously while, at the same time, improving their physical fitness. These so-called 'active commuters' mostly cycle on the 15 km-long high-quality bike path, that links the East Coast Park with the innercity area across the Marina Bridge.[14] The rise in transit that has been expanding in Singapore. Seniors are able to afford to travel on public buses, MRT and LRT at discounted senior fares,[15] cutting travelling time and promoting walking and cycling from their homes with sheltered walkways or dedicated cycling tracks to the stations or bus stops.

Over the years, Singapore's neighbourhoods and central business districts have been re-furbished, re-built or transformed. Becoming a functional walking city and those which have done this best have attracted the young talent and seniors at their prime of their careers to work and live in walkable locations such as the Raffles Place along Shenton Way/ Robinson Road/Cecil Street to the Tanjong Pagar and Anson subzones. There is a robust mandate in Singapore to offer an attractive place for seniors, both locals and foreigner, who are still fit in their *golden years* to contribute to its economy. In addition to promoting walking and cycling by having a well-equipped transportation system, seniors can benefit from easy access to nearby amenities and healthcare services. For instance, at the Jurong East MRT interchange, seniors can avoid the stress of road traffic and get some exercise by walking to nearby workplaces and shopping malls like Jem, Westgate and IMM Building. This also helps reduce carbon footprint and is environmentally friendly, as more seniors develop a habit of walking. Seniors find places like Jurong East appealing to live in, as they are just a few steps away from the Yuhua Village Market and Food Centre, and have access to the JurongHealth Campus, which boasts one

[14] Szubski (2022).

[15] Senior citizens of 60 years and above with concession cards can gain a discount of about 25% off adult card fares, but it is dependent on the travel distance. Seniors who travel pre-peak can also enjoy a 25% discount off the lower pre-peak fares. Senior Citizen Concession Pass at $60 (half the price of the Adult Monthly Travel Pass), permits unlimited travel on buses and trains, even in peak periods (Ministerial Committee on Ageing, n.d., -a, -b).

of the country's most advanced healthcare systems. The campus includes a 400-bed community hospital, the Jurong Community Hospital, and a 700-bed acute hospital, the Ng Teng Fong General Hospital. Overall, seniors in Singapore prefer to live in sustainable town centres, which is the third most important factor to consider when finding a suitable place of residence, after affordability and location. Seniors want to live in places where they can walk for most activities on a daily basis.

It can be argued that Singapore has made significant strides in becoming a city that caters to people of all age groups, ensuring a safe and nurturing environment for seniors to cherish their golden years while also setting a positive example for younger generations to follow.

Barrier-Free Accessibility for the Seniors in Singapore

Mobility and cognitive impairments are natural causes of ageing. Therefore, it is vital to enable the seniors to be independent and have access to the transportation and amenities as long as they could. By participating in civic and recreational activities, this can prevent loneliness or isolation, known to be the dominant reason of decline for most seniors. In 1990, the Code on Barrier-Free Accessibility (BFA) in buildings was enacted that incorporated universal design guidelines to cater to as many people as possible in public spaces. *Empathetic technology* sought to enhance the daily lives of the seniors through the DesignSingapore Council facilitating forums and publications, as well as collaborating with healthcare experts. In addition, the council also work with the private and social sectors to expand solutions for enhancing wellness for seniors.[16]

Ministerial Committee on Ageing Issues (CAI), under the Ministry of Community Development, Youth and Sports has spurred deeper harmonisation among the Building and Construction Authority (BCA), various town councils and LTA to provide uninterrupted connections when an individual walks and cycles from home to nearby neighbourhood amenities and public transit services. In 2008, the Transport Masterplan was launched by LTA followed by 2013 Masterplan. The primary aim is to provide a people-centred land transport network for an inclusive city.

[16] DesignSingapore Council, Ministry of Communications and Information (2015, October 1).

MRT stations are fitted with lifts, handrails and walkways. There are barrier-free access pathways with a fleet of wheelchair accessible buses that arrives frequently at bus stops as well as interchanges. The Green Man Plus programme is one of the measures and improvements that is put into operation. The elderly and the disabled would just need to tap the senior concession cards and they would get more walking time at traffic light crossroads.[17] A S$40 million accessibility fund was injected to facilitate a ten-year Barrier-Free Access (BFA) scheme from 2007 to 2016. One goal is to offer incentives to the private sector in order to uplift buildings constructed prior to 1990 which were not BFA compliant. A pilot project on City for All Ages was held in Marine Parade with 2,600 residents who voted that they desire for neighbourhoods and homes that are safe and makes life easy. Individuals from different age groups volunteered to walk separate neighbourhood lanes to identify possible hazards for the elderly. Following elderly-friendly improvements were instituted by LTA and the Marine Parade Town Council which include levelled void decks, larger block numbering, elderly-custom fitness corners and traffic crossings with more *green man* time.[18] To create a safer road environment, the elderly would have access to 50 Silver Zones Island wide by 2023.[19]

Transport for All was the underpinning theme for the Transport Masterplan 2040 which was launched in 2019. The masterplan outlines special needs and barrier-free in design and operational changes to promote inclusiveness. Hardware features were considered such as priority queues for seniors, handicapped and parents with prams as well as buses with wheelchair accessibility from 2020. Intangible improvements were incorporated including publicity campaigns to instil a gracious and caring commuting ethos and training bus drivers to handle these *special attention* passengers. This will enable seniors to travel independently, safely and confidently with the assurance of civicmindedness being a practice of commuters' daily living.

[17] Teo et al. (2006).

[18] D. Yeo (2013, January 7).

[19] Government of Singapore (2023, January 11).

Building a Liveable Social Infrastructure

Extant studies have recognised the mental and physical advantages of social connections, and it is evident there is an association between social connections, health and long-life expectancy.[20] The national socio-demographics suggest the likelihood of seniors living along and facing social isolation is real. Building a liveable social structure has always been on top of Singapore's mind in the nation's planning phases. There are laudable milestones that the island city-state has already make in this aspect. One key public body is the HDB which had made positive changes at the neighbourhood level. Amongst residential blocks, there are many *third places* for people to socially bond after work and outside their homes. This includes barbers and hairdressing salons, bird-singing spaces, childcare centres, grocery shops, hawker centres, inter-generational playgrounds and exercise areas, *kopitiams* (coffee shops), libraries, markets, schools and supermarkets. The lift lobbies and void decks of the housing blocks offer opportunities for people to exchange greetings and mix around while they are on their way back home from work and errands and while waiting to fetch their kids at the school drop-off areas. Grapevine can occur with chance encounters at the local *mom-and-pop* stalls, a common sight at every fourth HDB block. This was once a common sight in the past, but rare now, since many stallholders have retired or passed on. The circular rather than linear benches at the void decks in neighbourhoods were meant to promote communication and forge camaraderie among neighbours. Void deck spaces are utilised as much as possible and often fought over to provide social services to the community. These spaces should not go to waste since land is scarce in Singapore.

There are purpose-built premises like the Senior Activity Centres (SACs); a hub hive with activity for the seniors, offering seniors the opportunity to stay and mingle, connect, engage and offer support for one another. Many active ageing programmes for individuals and groups are run in SACs hand-in-hand with students and community groups, such as karaoke, craft and games. Such corporate social responsibility outreach programmes and community involvement programmes create opportunities for volunteers and seniors to engage meaningfully and learn from each other. For the older-old, these near-home spaces of the void decks, corridors and lift lobbies offer refuge with similar others, who

[20] Wahl and Lang (2003).

are likely less mobile or do not like programmed events. Since 2015, HDB has been planning and designing with the Ministry of Health (MOH) on horizontal co-location of public housing and nursing home projects. Tampines Greenweave was completed in 2019 and both sites were integrated with fenceless design. Clustering of facilities are instituted at the boundary of the two sites that encourages facility sharing and promotes engagement between residents in and outside the nursing homes. Example of these shared facilities are community living rooms and garden plots, elderly fitness equipment, exercise plazas, precinct pavilions and shelters, which all, would help to build a sense of community. The Senior Care Centres (SCCs) serve the senior residents staying close by and are conveniently located. SCCs provide care for frail seniors who need assistance with the daily needs (ambulation, dressing, feeding, toileting, washing), custodial care and rehabilitation (from occupational therapy, physiotherapy and speech therapy) and day-care activities include cognitive stimulation activities exercises, recreational activities like art and music and social activities, that aim to curb the weakening of mental and physical strength. Eligible seniors receive financial assistance of up to 80 per cent for day-care, transport and rehabilitation services. The Ministerial Committee on Ageing (MCA) announced plans in 2015 to launch at least ten Active Ageing Hubs (AAHs) amid the HDB estates. Targeting mainly the seniors, they can participate in many of the active ageing activities from basic health checks and talks, physical and social activities and care services to enable seniors to be mentally and socially engaged by being socially active. MOH plans to multiply the reach of SACs to support all seniors in the vicinity, no matter what their background or living arrangements. Along with centre-based services, a range of formal care services are available to support the seniors and their caregivers, these include befriending, integrated home and day-care services, medical escort and transport and Meal-on-Wheels.

The Singapore Government is reaching out to seniors living in private residential areas, including landed housing estates and condominiums. Over the past three years, cases of abuse, neglect and self-neglect have been on the rise, even among seniors who can afford to hire helpers for their daily needs.[21] In high-rise condominiums, other residents can avoid encountering seniors through private lift lobbies, which may suggest a

[21] The number of cases of abuse has risen from 232 in 2020 to 283 in 2021 and 338 in 2022, involving vulnerable seniors aged 65 years and above. See Ang (2022, April 4).

lack of consideration for seniors' desire for communal life and independence, regardless of whether they are mobile or not. Conducting private housing audits, in addition to public ones, would provide a better understanding of the living spaces seniors seek, such as barrier-free access to daily amenities and transit nodes. Seniors also hope to socialise with other residents who live nearby in communal spaces like neighbourhood community centres that are within walking distance.

Dementia Among Seniors and Healing with the *Kampong* Spirit

Dementia, on the contrary belief, is not a part of normal ageing nor is a specific disease. A common type of dementia is the Alzhemier's disease. It is evident that dementia affects the elderly who would experience impaired ability to think, remember, or decide on matters relating to their daily activities.[22] Due to its omnipresence in Singapore's ageing population, grassroots activists have initiated community awareness campaigns, so that residents at the neighbourhood level are aware of what dementia is and how this can impact one's life. Empathy can be shown for the dementia sufferers who can easily get agitated. They often wander around which can be disturbing to neighbours. The Agency for Integrated Care (AIC) has recruited businesses, services and residents. They form a Dementia Friends network working hand-in-hand in Dementia-Friendly Communities (DFCs). The aim is to be able to offer better support to individuals with dementia and their caregivers. Individuals with dementia and caregivers are supported with resources and support services such as Go-To Points. It operates as *safe return points* to bring together individuals with dementia with their caregivers when they are feeling helpless. AIC's idea thrives on the concept that neighbours and the community can play a vital role to enable seniors to *age-in-place*. An alert system is in place in the neighbourhoods. The emergency contacts of caregivers are in the local system so that they can be contacted at any possible time. For businesses which operate till the wee hours in the morning find a wandering individual that exhibits dementia symptoms, they can ensure

[22] See Division of Population Health, National Center for Chronic Disease Prevention and Health Promotion (2019, April 5).

the elderly is cared for in a safe environment until they are picked up by their caregivers.

Attention has been extended in urban planning and design so that environmental cues such as access controls, handrails and discernible street furniture can help the seniors in public spaces. By enhancing the legibility of the landscape, it would aid elderly residents in their wayfinding. The Building and Construction Authority of Singapore has introduced a universal manual on age-friendly design for the urban environment.[23] Outdoor design guidelines for persons with dementia have been established with major design elements around safety, ambience, technology, finishes and safety to enhance the urban experience for the seniors. The stipulated design aims to tackle disabilities related to declining motor cortex damage or cognitive skills. The ideas on a dementia-friendly design guidebook, *Looking to the Future*, was compiled in 2015 to recommend improvements that derive from the best practices on living environments such as the nursing homes. Six principles were outlined in terms of the dementia-friendly facilities, covering appropriate environmental stimulation, autonomy, empowerment, home-like environment, meaningful participation, personal space and privacy and safety, specially designed for the Singapore culture and context. In line with the kampong spirit, the dining areas are designed to accommodate four to six individuals, fostering social interaction while ensuring a comfortable experience for seniors with dementia, creating a welcoming atmosphere which reflects the unique dining practices and traditions that hold significance to Singaporeans. Beyond design guidelines, the built environment is supplemented with person-centred care—caregiving services that looked upon residents as unique individuals by embracing the standpoint of the person. The guidebook has helped over 2,000 stakeholders since its publication including staff from eldercare facilities, architects and policymakers. At present, stakeholders in Singapore's urban, social and healthcare segments persist to work together towards expanding inclusive designs to larger community spaces such as high-rise HDB neighbourhoods and parks.[24]

According to a study conducted by the Centre for Liveable Cities (CLC), it was found that a significant number of Singaporean seniors come from middle-income families, making them ineligible for subsidised

[23] Kua (2021).

[24] Fung (2015).

housing and care services. While lower-income seniors can access options like HDB rental housing paired with Senior Activity Centres and Senior Group Homes, these options are not available for middle-income seniors. On the other hand, commercial options managed by private companies, such as St. Bernadette's Home, are often too expensive. As a result, many middle-income seniors face challenges when seeking appropriate living arrangements and care. Some resort to premature admission into nursing homes or hiring foreign domestic workers, which may not align with their desire to age-in-place. Given that most seniors in Singapore reside in public housing, the key challenge lies in adapting to the high-density environment. The CLC study highlights that seniors prefer to live within their neighbourhoods and experience the vibrancy of their communities. They value their independence and engage in daily activities like socialising with neighbours, visiting local coffeeshops and markets and spending time with their grandchildren at the void decks. The aim is to create a supportive environment that allows seniors to maintain their independence while accessing necessary assisted services.[25]

In 2019, Ministry of Health (MOH), Ministry of National Development (MND) and HDB explored a new public housing assisted living concept to provide a less expensive option for integrated housing and care for the seniors. A pilot site was launched in Bukit Batok in early 2021. There are large common spaces on every floor, so intermingling among residents and their neighbours can happen as near as at their doorsteps, which allow seniors to age in a community of care and mutual support. Each housing units are fitted with senior-friendly design fixtures and spaces such as larger bathrooms to aid those who are less mobile. Besides that, a community manager is assigned to track the well-being of residents and they would have access to a basic service package that includes planned social activities at the communal spaces, 24 hours emergency response and special care dependent on their needs. Other discretionary services such as housekeeping, laundry and meals can be added onto the basic package if residents prefer so. The pilot project would inform the three agencies on areas to improve when considering expanding the public housing assisted living concept to other neighbourhoods. With the provision of additional eldercare option for Singapore seniors, they can age independently with self-worth, dignity and respect, while reducing

[25] See Centre for Liveable Cities (2017) study.

the premature admission into nursing facilities. The next step is to explore piloting the concept in the private market, in which the three agencies, MOH, MND and HDB have already plans in the pipeline.[26]

Singapore's Own Biophilic Society

A move to assist seniors includes deepening our understanding and rolling out effective mitigation measures from further physical and cognitive degeneration. In Singapore, psychiatrists, clinicians and increasingly social scientists research the connections between the brain, the body, a person's living environment and social atmosphere in correspond to a senior's welfare. The on-going Dementia Prevention Programme (DPP), which was launched in 2004, found the life expectancy and the quality of life could be improved for dementia sufferers together with stabilised health condition (for those with underlying medical conditions such as hypertension and diabetes) through meaningful lifestyle such as exercising, proper dieting and partaking in therapeutic activities including art, music rumination and horticulture.[27] Another milestone of the DPP, is that a pool of elderly volunteers have been trained under their pupillage, they arrange and tutor programmes for other seniors and in turn, increase their self-confidence when forging more social contacts. Research on preventive medicine *by*, *in* and *for* the community has been always on DPP's mind as the intention is to target depression prevention. Human beings are known to portray an intrinsic affinity towards forms of life and nature.[28] Seniors engaging in gardening is a way to enhance their mental health. Community gardens (Communityin-Bloom projects) are today clichéd in HDB neighbourhoods and have been successful in encouraging the seniors to step out from their homes. Pinning on the community gardens concept, NParks and National University Hospital's Department of Psychological Medicine designed a therapeutic garden at HortPark. The garden offers horticulture therapy to aid recovery for post-stoke patients, dementia sufferers and activity spaces to facilitate planned programmes.[29] More

[26] Kua (2021).

[27] Kua (2018).

[28] Biophilia, a concept coined by entomologist Edward Wilson, refers to "the innately emotional affiliation of human beings to other living organisms. Innate means hereditary and hence part of ultimate human nature". See Wilson (1984, p. 31).

[29] See Ministerial Committee on Ageing (n.d., -a, -b).

of similar therapeutic zones are scheduled for health-related facilities and existing parks. The horticultural therapy designed for the treatment group consists of a weekly session for months 1–3, then a monthly session for months 4–6. Seniors get to learn planting vegetables and pick up gardening skills such as transplanting and composting. The participants will get to prepare a soup dish after harvesting their full-grown vegetables in a collective effort. Biomarker results show for the group who receive treatment compared to the control group, the former experienced a large drop in a pro-inflammatory protein (cytokine interleukin-6 or IL-6). This means they demonstrated a better immune response than the control group. The treatment group also fared better in their memory, psychological well-being and life satisfaction scoring compared to the other group, showing positive outcomes from the horticulture therapy. It was found that seniors from the treatment group cope better because of the opportunities they have to be able to forge social connections and build trusting relationships with others.

Efforts to expand the value of horticultural therapy so that more seniors can enjoy its positive effects seemed practical and points to initiatives of intensifying the biophilic city concept within Singapore's urban environment. Pursuing biophilia brings the society. Psychosocial interventions have shown to have a positive impact on the mental health of seniors, promoting beneficial changes in their behaviour, such as fostering new friendships within their local community. Another finding is that the old-old (75 years and above) who are supported by the young-old (65–74 years), who are feebler, were found to have more compassion. This form of value change signifies that the biophilic city concept can be rooted in and by summoning together the capacities in the community, these initiatives can be executed at the individual and small-unit levels. Given the resources to promote multiplier impacts at a broader level, planning for successful ageing widens to become a societal undertaking so Singaporeans is a country that is not just liveable and thriving, but also kind, generous and gracious.[30]

[30] For details, see Kua (2018, 2021).

Final Thoughts

Singapore's population is ageing. One in four Singaporeans will be aged 65 and above by 2030. Changing family structures with more seniors staying alone, a shrinking workforce and increased healthcare needs are some of the emerging new realities of the city-state. If Singapore plans well for these changes and seniors lead meaningful lives, the nation can gain the silver dividends that comes along with increased longevity. An Urban Systems Study marks Singapore's journey in planning for an ageing population, documenting the initiatives implemented from the 1980s. To ensure that planning continues to strive in satisfying seniors' needs and interests, the government boarded on a year-long consultation process in 2014 comprising some 4,000 residents from different walks of life, to gauge their aspirations for ageing. A national blueprint that encompasses the Action Plan for Successful Ageing including public spaces, transport, employment, health and wellness to make the country a *City for All Ages*.

Transcending to prepare for a future-ready city, as Singapore transforms itself, the debate on how and the extent to which, as to where, the government intervention is still required, is a continuing dialogue among government bodies and beyond. There is a need to have clear routes to deal with short-term and longer-term ideals and challenges. While there is a growing trend of social networks and community heartware assuming deeper roles to care for the seniors, the direct caregivers also need support to look after their aged family member. In the recent years, there are wider spread programmes for the seniors which include dementia-friendly neighbourhoods and assisted living, other than a complementary range of housing and care choices that form a part of the urban landscape. The still-mobile seniors can opt walking amid the HDB void decks and around their neighbourhoods or catch-up with their neighbours for a good ***kopitiam*** (coffeeshop) session. The accessibility of the iconic therapeutic HortPark and programmes for seniors had flourish the biophilic city concept within Singapore's urban environment. The young-old (65–74 years) supporting the old-old (75 years and above) not only lends an extra arm of manpower but also widens societal efforts to enable more seniors age gracefully. Voluntary commitments rendered by GoodLife! which is an activity centre that was setup by the Marina Parade Family Service Centre and Agency for Integrated Care (AIC) that primarily support seniors with dementia can be linked to SDG 17—Partnerships for Goals.

Even though policy frameworks are principally in place, more needs to be done on how best to put into action from a human-centric approach. Combating ageism and planning for its ageing population will always be certainly on the cards for Singapore. It is essential to rally community partners, corporations and individuals from various backgrounds in Singapore to establish a robust support and care network for the increasing population of seniors. This challenge requires the collective effort of residents, extending beyond the responsibility of the government alone. When no one gets left behind, then truly, Singapore can be coined as a *city for all ages*, while the nation strives towards becoming a gracious, ageless and liveable nation. By implementing the measures discussed, seniors in Singapore can maintain good health and well-being, aligning with the Sustainable Development Goal 3—Good Health and Well-Being.

References

Agency for Integrated Care Pte Ltd. (2022). *What is dementia?* AIC Dementia. Accessed on 8 May 2022.

Ang, S. (2022, April 4). *Elder abuse on the rise, but no increase in higher-risk cases: Sun Xueling*. https://www.straitstimes.com/singapore/politics/elder-abuse-on-the-rise-but-no-increase-in-higher-risk-cases-sun-xueling. Accessed on 21 March 2023.

Centre for Liveable Cities. (2017). *Piloting assisted living: Research and workshop insights*.

Cheng, T. C., Li, J., & Vaithianathan, R. (2019). Monthly spending dynamics of the elderly following a health shock: Evidence from Singapore. *Health economics, 28*(1), 23–43.

Chong, K. H., Jia, Z., Loo, D., & Cho, M. (2015). Successful aging in high-density city state: A review of Singapore's aging policies and urban initiatives. In F. G. Caro, & K. G. Fitzgerald (Eds.), *International perspectives on age friendly cities*. Routledge.

DesignSingapore Council, Ministry of Communications and Information. (2015, October 1). *Design to improve lives of the elderly*. https://www.nas.gov.sg/archivesonline/data/pdfdoc/20151008002/media_release_-_aidi_publications_launch_final.pdf. Accessed on 8 May 2022.

Division of Population Health, National Center for Chronic Disease Prevention and Health Promotion. (2019, April 5). *What is dementia?* https://www.cdc.gov/aging/dementia/index.html#:~:text=Dementia%20is%20not%20a%20specific,a%20part%20of%20normal%20aging. Accessed on 14 May 2022.

Fung, J. C. (2015). *Dementia design sourcebook: Design guide, design elements*. Department of Architecture, National University of Singapore.

Gee, C., Arivalgan, Y., & Chao, F. (2018). *Singapore perspectives 2018: Together*. World Scientific Publishing Corporation.

Government of Singapore. (2022, April 8). *Electric vehicles*. https://www.lta.gov.sg/content/ltagov/en/industry_innovations/technologies/electric_vehicles.html?gclid=Cj0KCQiAo-yfBhD_ARIsANr56g736-zSBBVL3eDJSad2w4OeZOmHfCJhpLRJ0VHM32issr6OI63nHVIaAlc2EALw_wcB (Accessed on 26 February 2023).

Government of Singapore (2023, January 11). *Certificate of Entitlement (COE)*. https://onemotoring.lta.gov.sg/content/onemotoring/home/buying/upfront-vehicle-costs/certificate-of-entitlement--coe-.html#:~:text=Understanding%20COE,-All%20vehicles%20in&text=To%20register%20a%20vehicle%2C%20you%20must%20first%20place%20a%20bid,exercises%20conducted%20twice%20a%20month. Accessed on 26 February 2023.

Koh, W. T. H., & Lee, D. K. C. (1994). The vehicle quota system in Singapore: An assessment. Transportation research Part A: Policy and practice. *Research Collection Lee Kong Chian School Of Business*. *28*(1), 31–47. https://ink.library.smu.edu.sg/cgi/viewcontent.cgi?referer=&httpsredir=1&article=4363&context=lkcsb_research#:~:text=The%20first%20COE%20auction%20was,auctions%20on%20a%20quarterly%20basis. Accessed on 26 February 2023.

Kua, E. H. (2018). *Colours of ageing: 30 years of research on the mental health of the Singapore elderly*. Write Editions.

Kua, E. H. (2021). *Towards ageing well: Planning a future-ready Singapore*. Centre for Liveable Cities Singapore. https://www.clc.gov.sg/docs/default-source/urban-systems-studies/uss-towards-ageing-well.pdf. Accessed on 8 May 2022.

Ministerial Committee on Ageing. (n.d.-a). *What is the action plan about? I feel young SG—MOH*. https://www.moh.gov.sg/ifeelyoungsg/about/what-is-the-action-plan-about. Accessed on 8 May 2022.

Ministerial Committee on Ageing (n.d.-b). *Go about easily*. https://www.moh.gov.sg/ifeelyoungsg/how-can-i-age-in-place/go-about-easily#:~:text=A%20SENIOR%2DFRIENDLY%20TRANSPORT%20NETWORK&text=Senior%20citizens%20with%20concession%20cards,the%20lower%20pre%2Dpeak%20fares. Accessed on 26 February 2023

Newman, P., & Kenworthy, J. (2015). *The end of automobile dependence* (pp. 105–140). Island Press/Center for Resource Economics.

Singapore Department of Statistics. (2019). *Population trends 2019*. https://www.singstat.gov.sg/-/media/files/publications/population/population2019.pdf. Accessed on 8 May 2022.

Singapore Department of Statistics. (2021). *Population trends 2021*. https://www.singstat.gov.sg/-/media/files/publications/population/population2021.ashx. Accessed on 8 May 2022.

Szubski, C. (2022). *Singapore: Cycling*. https://sportifycities.com/singapore-cycling/. Accessed on 26 February 2023.

Teo, P., Mehta, K., Thang, L. L., & Chan, A. (2006). *Ageing in Singapore: Service needs and the state*. Routledge.

Wahl, H. W., & Lang, F. R. (2003). Aging in context across the adult life course: Integrating physical and social environmental research perspectives. *Annual Review of Gerontology and Geriatrics, 23*, 1–33.

Wilson, E. (1984). *Biophilia*. Harvard University Press.

Worldometer. (n.d.). *Life expectancy of the world population*. https://www.worldometers.info/demographics/life-expectancy/. Accessed on 8 May 2022.

Yeo, D. (2013, January 7). *Making Singapore a city for all ages*. Challenge. https://www.psd.gov.sg/challenge/ideas/deep-dive/making-singapore-a-city-for-all-ages. Accessed on 11 May 2022.

CHAPTER 12

Biophilic Vision, Regenerative Sustainable Urbanism and Circular Economy

The Paris Agreement (COP21) emerged as a significant global response to address the risks associated with climate change by emphasising the urgent need for widespread decarbonization.[1] While various large-scale geoengineering initiatives have been proposed, an alternative and potentially more effective solution could be found within the cities themselves. Through the concept of city bioregions and the implementation of green urbanism, cities can serve as viable substitutes for traditional geoengineering projects while offering comparable benefits and outcomes. The combined approach is deemed less risky and has multiple, added social and economic advantages, other than reducing urban ecological footprint. However, achieving success in this endeavour hinges on the ability of city

[1] The Paris Agreement was sealed at the UN Climate Change Conference (COP21) in Paris, France, on 12 December 2015. An international treaty among 196 parties on climate change, the agreement came into force on 4 November 2016. World leaders aimed to limit global warming to 1.5 °C by the end of twenty-first century to mitigate the drastic climate change such as severe and more frequent heatwaves, drought and rainfall. The plan is that the greenhouse gas emissions must peak before 2025 and decline 43% by 2030. Since 2020, more countries and cities have made national climate action plans to reduce their greenhouse gas emissions (decarbonisation) and establish carbon neutrality targets, most noticeable in the power and transport sectors, in order to reach the Paris Agreement goals.

Melissa Liow Li Sa and Sam Choon-Yin, *Sustainable Urban Development in Singapore*, Sustainable Development Goals Series,
https://doi.org/10.1007/978-981-99-5451-3_12

makers and policy writers to operate at a large scale and meet rigorous standards of urban sustainability performance.

There are three horizons of green urbanism. The first horizon, green design is to improve traditional urban development. The second, sustainable development, often coined as the preliminary move towards a net-zero impact. The third is regenerative urbanism, an evolving concept that empowers biosphere restoration. Cities such as Singapore have adopted green urbanism due to the call to do its part to mitigate climate change challenges. Singapore has been dedicated to becoming a sustainable city, with a longstanding commitment to the Garden City Concept since May 11, 1967. The Singapore Green Plan 2030 aims to transform the city into a City in Nature, with a focus on conserving and expanding Singapore's natural capital and improving the health and well-being of its residents. The nation also harnesses design and technology including the biophilic vision to boost urban metabolism and offer positive sustainability performance. Singapore's orthodox approach to regenerative urbanism serves as a primary urban geoengineering push, while concurrently presenting life-affirming co-benefits to this nascent city. This should not take googol years amid the entropy of regenerative of urbanism which is already encrypted in Singapore's DNA. Serious attention and efforts are already underway in this city-state, not forgetting the Singapore city dwellers embracing more walking and cycling in their daily lives to address climate change challenges.

Biophilic Vision of Singapore

Singapore has become a first world and cosmopolitan city. As the nation transforms, there is more drive to greening the city-state, raising it to a high priority government issue. NParks was tasked to implement the city visions—*Garden city* in 1963, then, *City in a Garden* City, *City in Nature* and *Singapore Green Plan 2030* to brand Singapore a *liveable city*. NParks was later subsumed to be a part of the Ministry of National Development. The combined efforts of NParks, URA and HDB cover diverse work scopes and the territory in entirety. Seeking an equilibrium between conservation and development has been a continual endeavour by the three public agencies. Despite the loops and hurdles, they have realised the increased of vegetation cover that had grown from 36 to

47% between the 1980s and 2012 under their management.[2] Given its small land area of 700 square kilometres, Singapore faced the challenge of avoiding detrimental economic compromises while providing enough space for its growing population. Despite this, the country managed to allocate 9% of its land area to nature reserves and parks, promoting walking and cycling as means of transportation.[3] These green spaces not only contribute to the health of Singapore's population but also align with Sustainable Development Goal (SDG) 3—Good Health and Well-Being, as they enable people to breathe clean and fresh air during their daily activities.

The biophilic vision of Singapore was first conceptualised in the 1960s. Then Prime Minister Lee Kuan Yew put forward the design of a green and clean environment to alleviate the stark concrete urban environment and enhance the quality of life of its city dwellers, henceforth, charting Singapore's journey into becoming a *Garden City*. Singapore's green and clean environment has been able to fulfil the recreational and lifestyle needs of the burgeoning affluent society who can opt to walk, cycle and ride on public transport to their work and residential places. The nation's appeal as a place for international talents and businesses, transforms Singapore into a vibrant global city. The idea of a City in a Garden was formulated with a vision to engage the community and implement more advanced strategies for preserving Singapore's natural heritage through greenery plans. The goal is that Singapore would be recognised as a bustling metropolis nuzzled in a flourishing shroud of tropical greenery.

According to a recent study conducted by The Network and Boston Consulting Group (BCG), Singapore has emerged as one of the most appealing destinations for top-tier foreign talent. The city-state has jumped 10 spots to secure the 8th position in the list of the most attractive locations for global talent. This can be attributed to Singapore's reputation as a highly liveable city with ample public spaces, low crime rates and exceptional healthcare systems. These factors make Singapore an ideal destination for expats and their families. By focusing on sustainable economic growth and creating full and productive employment opportunities, Singapore is able to align with SDG 8—Decent Work and Economic Growth.

[2] Newman (2014).

[3] Civil Service College Singapore (2008, June 1).

Singapore gradually built its standing as a *City in Nature*—the concept is an extended vision of the theme *Garden City* back in 1960s. Amidst the Singaporean blueprints, there has been a fervent awareness and deep-seated consciousness of green spaces, and urban planners have intertwined nature in buildings amass all over the city. At present, all new developments include some form of plant life, be it verdant walls, green roofs, or cascading vertical gardens. Its tree canopy reached close to 30%, and half of the nation's land is covered in green space. The parks offer exercise space and more breathing air, otherwise act as the *green lungs* amidst the compact city environment. There are existing projects and those in the pipeline that are multifunctional in supplying ecosystems services, people amenities and transpiring the likelihood to discover the *wilder nature* side of Singapore where people can walk and cycle to reach their destinations.[4]

City greening was no easy feat for Singapore. It took the implementation of several innovative technologies with a focus to improving its urban biodiversity. NParks opened a National Biodiversity Centre in partnership with the United Nations Convention on Biological Diversity and released Singapore's own Biodiversity Index to measure urban biodiversity performance globally.[5] The nation has mottoes for development and conservation in its administration, legislation and economic structure. Despite so, the primary tensions impacting biodiversity in Singapore arise from scarce habitat availability, land use and adverse climate changes. Master planning, legislation and education are critical obligations to address the population and land scarcity issues. Various alleviation stratagems are implemented to mitigate the climate change challenges. A few noteworthy projects include forming frontline barricades, establishment of nature reserves aiming to conserve, protect and manage remaining native species and ecosystems, expansion and preservation of mangroves and forests on land, coral reefs at sea and widespread use of greeneries on and between buildings structures. The island city-state is the host to four natural reserves—Bukit Timah, Central Catchment, Labrador and Sungei Buloh and two national

[4] Leonard Ng, the County Director at Ramboll Studio Dreiseitl, likened Singapore to the canary in the coal mine when it comes to biophilic design. Ramboll Studio Dreiseitl has been involved in several remarkable parks and biodiversity projects, one of which is their contribution to the Bishan-Ang Mo Kio Park that offers a unique and close connection with nature. DesignSingapore Council (n.d.).

[5] United Nations-Habitat (2013).

parks—Fort Canning Park and the Botanic Gardens. In addition, Singapore has varied nature spaces—629 community gardens, 223 heritage tress, 415 public parks, 322 skyrise greeneries and 5 heritage roads. To realise the demarcation of these dedicated nature spaces, NParks was active in facilitating inter-agency collaboration, public use administration, protection, conservation and management of ecosystems, habitat restoration and species loss management.[6]

Beyond the Singapore Green Plan 2012, there is a generation of new niches of greeneries that are sprucing up on and between buildings structures.[7] Walls, roofs, sky terraces and landscaped balconies become living construction masterpieces and multistorey greeneries. New broadening horizons for greening the city-state include its vacant spaces, streetscapes, roads, green plots and heritage trees. Many areas were restored by NParks by introducing new parks that sit on spaces that were formerly disrupted by development. Nature reserves and parks are more accessible to residents with the planning and implementation of 360 kilometres park connectors all-round the island-state. A walk in the park is not just another form of exercise like walking down a city street, which puts demand on attention like navigating through crowds and dodging cars. It puts less demand on attention thereby a more effective way to replenish energy, restore memory and improve cognition.[8]

Other highlights comprise of the Reforestation and Outreach Programme and the HSBC Tree Top Walk which is popular for a free-standing suspension bridge amass the MacRitchie Reservoir Park within the Central Catchment Reserve.[9] Hitherto was land composed by road, railway tracks, foreshores and drainage reserves before the direct connectors made their way into the new landscape. One of the nostalgic natural and historical attractions is the North–South Green Corridor alongside

[6] NParks has implemented urgent measures to address climate change and its effects, which are seen as contributing significantly to the attainment of SDG 13, which concerns Climate Action (Chan et al., 2010).

[7] Newman (2014).

[8] Elizabeth Gosling and Kathryn Williams found a relationship between connectedness to nature and the tendency to care for the environment. For example, farmers in northwest Victoria, Australia, who were more connected to nature, were more likely to manage their farms in an environmentally friendly way, suggesting that emotional association with nature leads to an expanded ecological sense of self and pro-environment behaviour. See Gosling and Williams (2010).

[9] Chew (2016).

the *Keretapi Tanah Melayu* (Train of the Land of Malaya) railway from Woodlands, close to the Straits of Johor passing through Chinatown in the north then ending at the 1932 built art-deco station. The attraction was opened to the public in 2011 after ceasing service a century before. Many recreational activities can be enjoyed by residents and tourists such as camping, hiking, skating, walking, cycling and mountain biking. Avid nature lovers can do plant, animal and nature watching in forest canopies, learn to appreciate, conserve and identify flora and fauna, as well as facilitate research surveys that help NParks to comprehend better how forest ecosystems work.[10]

According to NParks and URA, Singapore had added 30 hectares of sky or high-rise greeneries between 2019 and 2012 and aims to reach 50 hectares of skyrise greeneries on buildings by 2050. With the addition of greenery in Singapore's skyline, walking or cycling from one's workplace or doorstep to another destination becomes more enticing. The *City in a Garden* vision is partly achieved with this strategic move. This special greening technique would solve limited land issues, improve air conditions and address the Urban Heat Island (UHI) effects, and as a result, promote sustainable urban development. NParks carries out research, develops guidelines and incentives for the implementation and promotion of skyrise greenery through publications, seminars and awards for vertical greenery, green roofs and gardens in the sky. This innovative technology was coined in the early twenty-first century. The aim of the national Skyrise Greening initiative is to perpetuate the high-rise greenery concept to health and education facilities, office and commercial offices and residential towns. Policies aiming to speed up a radically new approach to bringing nature into the Singapore city have never been more apparent than in its endless modernist, cookie-cutter high-rise towers. There are the GFA Exemption for Communal Sky Terraces, Landscape Replacement Policy for Strategic Areas, Skyrise Greenery Incentive Scheme and Outdoor Refreshment Areas on Landscaped Roof Tops.[11]

The world's first biophilic hospital was the Khoo Teck Puat (KTP) Hospital established in Singapore in 2010. The hospital has 550 beds in public and private wards, and many biophilic features that are observable

[10] NParks Rail Corridor (2023, April 3).

[11] Yok (2012).

from its building architecture. There is a community garden run by volunteers, green walls, vegetated courtyards with ponds, multilevel gardens and a green roof that locates an edible garden that supplies fruits and vegetables for the hospital's kitchen.[12] Findings spell from the growing body of research on how the biophilic concepts benefit in improving health and well-being of patients suffering from chronic illnesses. Patients experience less pain and speedier recovery after surgery. Greenery skyrise and its innovative technologies are found to soothe the UHI and climate adverse effects. Acting as a heat insulator and rainfall absorber, these technologies are able to reduce flood risks, energy consumption and maintenance costs.[13]

HDB formulated the Landscape Guide that offers technical and design support for installation of skyrise greenery. The goal is to understand the role of landscape design as a provider of community spaces that led to the following notion—*Community-friendly*, *Environmentally sensitive* and *Aesthetically pleasing*.[14] The Landscape Guide customs design rooftop greenery in HDB estates to suit the different architecture taxonomies to improve environmental, aesthetic and thermal performance. The landscaped decks link the level difference of adjoining sites. People can relax at the seating or bench corners in the courtyards while admiring the think foliage or organise a catch-up session with friends. Rooftop gardens located on the top of the multistorey car park offer extended landscaped area for recreational activities such as walking, cycling, strolling, exercising, playing and hold casual meetings. The rise of greenery skyrises in the Singapore city is a local phenomenon nowadays, which can be attributed to URA's painstaking efforts. URA has had a seismic hand in the recent high-rise residential developments like the fifty-storey Pinnacle@Duxton. It has a semi-open space that offers shelter from the sun on its 26th floor. It would be suitable for day-time activities and has a landscaped jogging trail. On the 50th floor, there is a roof garden and a landscaped roof on top of the car park. Another Singapore's remarkable design innovations can be observed in Punggol Breeze. It hosts a seamless access to a multistorey car park and a full-equipped rooftop garden for residents to go outdoors and partake in recreational activities.

[12] Newman (2014).

[13] Auger (2013).

[14] Housing and Development Board (2013).

Dawn of New Greening Schemes

As part of the 2014 Master Plan, URA developed the Natural Green and Blue Master Plan. The aim is to enhance the development and preservation of waterways and new green areas that offer convenient public accessibility, rhetorical education about the new greening schemes and recreational spaces. Design guidelines are shaped with the goal to protect the remaining environments and ecosystems of the island city-state. Examples include the *Sungei Buloh* Wetland Reserve which comprises additional 38 hectares of new wetland park with walking trails added on. The area would be appealing for experiential and discovery learning to take place while protecting its inhabitants for biodiversity all together. As for the Northern Explorer Park connector, a 25-kilometre walking stretch will be added on, and the north region will increase by another 200 hectares to its present 700-hectare of parks.

The emphasis of the current vision—*City in a Garden* is to mitigate climate change effects and urban sustainability. Rejuvenation of streetscapes and revamping urban parks are efforts to realise the vision and encouraging walking and cycling for short distances. Parks are now multi-functional green spaces that fulfil social and environmental purposes. In addition, parks are biodiversity enrichment spaces, carbon sinks and pilot testing grounds to experiment sustainable urban features to address tropical climate challenges such as fast evaporation, high temperatures and heavy rainfall. New parks foster recreational activities, social gatherings and cultural events. This results a shift of our mindset of parks and its role in enabling natural and human systems co-development. URA has identified three destination parks as part of its planning initiative—the East Coast Park located at the southeast of Singapore, the Jurong Lakes Park in the southwestern territory, and the Admiralty Park in the north. Bodies such as URA optimise the use of these destination parks by carrying out experiments of different sustainable urban features suited for tropical climate countries such as Singapore. Public are attracted by its 9-kilometre chain of open green spaces, named the Southern Ridges connecting hill parks with two elevated bridges, boardwalks and trails. Visitors can enjoy walking and cycling while observing without disturbing the flora and fauna in its own habitats, promoting protection and appreciation of these nature sanctuaries.[15]

[15] Urban Redevelopment Authority (2022, May 18).

The other archetypal example is the eco-nature neighbourhood park known as Greenwood Sanctuary at Admiralty. The park hosts greeneries to recover from microclimatic conditions, contains drainage and land forming systems, acts as an added green lung for compact built environments, and easy maintenance within HDB housing towns.[16]

The City in a Garden concept in Singapore has proven to be effective in achieving several Sustainable Development Goals (SDGs). By prioritising greenery and biodiversity, Singapore has taken significant steps towards protecting life below water (SDG 14) and life on land (SDG 15). Additionally, Singapore's efforts to combat climate change and its impacts through sustainable development align with SDG 13—Climate Action. It is evident that incorporating nature and greenery into urban planning can have a positive impact on multiple SDGs and contribute towards creating a more sustainable and resilient future. Another awe-inspiring part of the *City in a Garden* vision is the ABC (Active, Beautiful and Clean) Waters programme. Spearheaded by PUB in 2006, the programme hopes to uncover the hidden potentials of Singapore's reservoirs and waterways. The main objectives of this transformation are to integrate water infrastructure into the urban fabric, serving as a central water channel for communities and providing a serene urban and green landscape for pedestrians and cyclists along its waterways. The Kallang River at Bishan-Ang Mo Kio Park is a flagship project under this programme. Situated in the Central catchment, it has several functions from improving the site aesthetics, using special plants to purify the Kallang River water, promote contact, connectivity, sense of ownership, organising education and recreational activities for residents between the parks and housing estates. The project restored once more the sparkling rivers with landscaped banks and restoration of original vegetation and return of insects, fish and otters living amidst these postcard-pretty community spaces. The Marine Barrage is a water infrastructure of the urban landscape which act as flood controls, water reservoir and offer many recreational activities for the community.[17]

[16] Ma and Guek (2012).

[17] PUB (2022).

Three Horizons of Singapore's Urban Sustainability Performance

There are three horizons of green urbanism commitment: *green design*, *sustainable development* and *regenerative urbanism*. For every horizon, it requires a stage-change in urban planning innovation and system assimilation. Green design innovations produce lower ecological footprint. Sustainable urban development means achieving net zero of footprints. Regenerative cities deliver net positive outcomes through circular metabolism.[18] In the case of Singapore, simple green design methodologies are now common, sustainable development is somewhat practised and regenerative design is starting to emerge. The challenge is that there is no clear distinction when determining the nation's urban sustainability performance and its effects. There is an oversupply of names that Singapore has under its belt. Scholars often used terms such as *low-carbon city*, *sustainable city*, *smart city*, *eco city* and *resilient city* to describe cities such as Singapore that had made remarkable progress in green urbanism initiatives. What needs more attention is to establish a consistent definition of performance metrics while applying an urban metabolism model.[19] From the city planning perspective, planners need to comprehend the comparative possibility of an identified space to be converted and to improve weaknesses, whether the pace of development is green, sustainable, or regenerative. Singapore's progressive green design and development meets SDG 13—Climate Action while making Singapore city safe, resilient, sustainable and inclusive, fulfils SDG 11—Sustainable Cities and Communities.

Singapore has made considerable efforts to recover from the aggressive industrialisation era in the 1970s, 1980s and 1990s, which included initiatives such as artificial coral breeding, restoration of vegetation and ABC Water Programmes. These programmes were implemented with the goal of preserving life below water (SDG 14) and life on land (SDG 15). The restoration of vegetation not only aids in preserving life on land but also serves as carbon sinks, supporting efforts towards SDG 13—Climate Action. Conscious efforts in restoring degraded environment and

[18] Newton (2007).

[19] Broto et al. (2012), Baccini and Brunner (2012) and De Jong et al. (2015).

improving the biosphere are made while urban system integration is instituted to fulfil human needs. Such efforts seek to repair and regenerate sustainability performance measurements linked with planetary frontiers at the local scale, for instance the water, nutrient cycles, greenhouse gas (GHG) emissions across every stage of development. At its most effective, Singapore in its endeavour to partake in regenerative urbanism will have urban infrastructure that results in less waste and more waste recovery, reduced water consumption, more local food production and cleaner energy. The aim is to project sustainable energy consumption and production patterns (SDG 12).

The three urban horizons provide insights into decarbonizing cities like Singapore and assessing their overall urban performance at different scales, including city, district, or individual plot levels. Sustainable development aims to maintain a steady state of core resources, preserving biodiversity and achieving net-zero emissions. This steady state acts as a baseline, and any performance below it indicates the need for urban sustainability upgrades or green design interventions. Conversely, performance above the baseline signifies regenerative practices that restore nature. Green design focuses on minimising negative impacts compared to traditional development, but still involves some degree of biosphere degradation. Regenerative urbanism goes further by actively restoring nature and requires achieving a net-negative carbon dioxide equivalence, either through offsets or on-site reductions. Calculating carbon dioxide equivalence involves complex methods that adhere to carbon accounting standards such as GHG protocols. To optimise efficiencies, it is beneficial to integrate on-site, integral and district-wide infrastructure at the city scale rather than focusing solely on individual plots. Shared infrastructure and distributed renewable energy supply can generate economies of scale, particularly when covering larger areas like a city or district. Higher population density can lead to lower costs per dwelling and greater energy savings through improved appliance efficiency, transportation efficiency and heating and cooling efficiency. This enables the provision of affordable and clean energy, aligning with SDG 7.[20]

[20] See Beatley and Wheeler (2004) and Newton et al. (2012).

Singapore Buzz on Circular Economy

There is an increasing need to dispose the solid waste and sewerage that comes with its growing economy, water and energy consumption. The nation faces a lack of these public utilities, and particularly so for Singapore since land and resources are limited. It would be very soon that the Semakau Landfill, the nation's only landfill will be loaded up within years based on the current level of waste production. In order to grow sustainably, Singapore is driving to overcome these challenges by embracing a circular economy approach. A shift seemed seismic that is to change an entrenched mindset from *use and throw* mindset to *reusing* resources for a long time.

What is a circular economy? The circular economy is an alternate way to the conventional linear economy that consumers *take, make and waste.* It calls for recovery of resources at the product's end life by feeding back into the production cycles. While creating new growth opportunities, it can address waste management issues, the burdens on the environment of the nation's usage and production, and even drive better resource productivity and a level of competitiveness in the global economy. Singapore has implemented policies to foster sustainable production and usage and support efforts towards responsible consumption and production (SDG 12). In 2021, Extended Producer Responsibility (EPR) was launched, starting with e-waste. Producers are liable for the *end of live* of their products. Due to the familiarity of their own products, they are more motivated to recycle, as well as designing new products and creating innovative circular business models. Eventually a 360-degree closure of their waste loop is formed. Additionally, the 'Closing the Waste Loop' initiative which is a research grant call for sustainable design of plastic materials was implemented. This constitutes a research alliance of research institutes, higher education institutes and businesses to explore novel methods to capture innovation value and address the pertinent plastic problems. We are unable to remove plastic totally from our lives, but we can change our approach. This initiative is aimed so that plastic becomes easier to recycle while able to obtain value from the waste plastic that is reusable.[21]

The circular economy begins from charting a route to becoming a Zero Waste Nation in Singapore. A Zero Waste Masterplan was launched in 2019 which is one of the city-state's programmes which was presented at

[21] Open Government Products (2022, April 19).

the Urban 20 (U20) Mayors' Summit. The COVID-19 pandemic has led to less production and carbon footprint due to decline in global demand for products. It has stimulated the development of a carbon–neutral and circular economy. One area of focus is to expand effort in recycling and reducing waste streams. Singapore had managed to recycle up to 99% of its construction and demolition waste in 2018 for the nation's construction industry. Endeavours are on-going in the other three primary waste streams: e-waste, packaging waste and food waste. The aim is to produce similar success stories. A RENEW programme was rolled out in 400 locations island-wide with over 450 bins (as of March 2019) to increase ease for recycling E (or electronic) waste in joint efforts with DHL, Starhub and TES-AMM. In 2021, the National Environment Agency (NEA) expanded the scope of the Extended Producer Responsibility Framework initiative. Under this mandate, producers are now required to take physical and financial responsibility for the management of their end-of-life products. This includes various electronic waste (e-waste) products such as batteries, lamps, solar panels, ICT equipment and large electronic and electrical appliances. The Singapore Packaging Agreement, implemented in 2007, aimed to tackle packaging waste by implementing strategies such as consumer education, supply chain initiatives and the redesigning of product packaging and production processes. This agreement aimed to promote waste reduction and sustainability throughout the packaging industry. NEA worked with grassroot bodies, non-governmental organisations and businesses to motivate residents to bring their own shopping bag. NEA is also looking at transforming plastics into solid recovered fuel through a pilot project that harness the mechanical biological treatment methods.

Food waste treatment has seen some progress too. Food waste treatment has made significant progress in Singapore. Public housing estates, such as Tampines GreenLace, have piloted food waste recycling initiatives, with one in three households using recycling bins on a weekly basis. Food centres and shopping malls have implemented separate food waste collection and on-site treatment processes. This treatment converts food waste into non-potable water for various uses and produces compost or liquid nutrient for landscaping purposes. Nutrient recycling methods, such as converting soya bean waste and spent grains into animal feed, are projected to meet 30% of the local nutritional needs by 2030. For food waste that cannot be treated or recycled, there are Waste-to-Energy plants where it is discarded for energy recovery. A notable project in this regard is the treatment plant at Tuas Nexus, which utilises co-digestion of food waste and sludge from wastewater treatment plants. This process has yielded more biogas and resulted in cost savings by reducing over 200 thousand tonnes of carbon emissions per year. In addition to food waste, efforts are also being made to recycle horticulture waste. Gardens by the Bay and other parks in Singapore recycle approximately 70% of horticulture waste, including plant trimmings and tree branches. The horticulture waste is burned in an on-site biomass power plant located at Gardens by the Bay, generating energy to power the chillers of the Cloud Forest and the Gardens' Flower Dome. This approach provides a sustainable and affordable source of clean energy.

Circular Economy Changing Businesses in Singapore

Singapore has been persistently striving towards transforming into a sustainable city, with various possibilities for exploring new innovative designs, business models and supply chain models through the growing circular economy. The advantages of the circular economy are evident in reducing the nation's reliance on new materials and natural resources. However, the crucial question remains, how can businesses apply circular economy principles to increase revenue and become more sustainable?

Here is a list of recommendations suggested by The Explorer (March 12, 2020) that Singapore can further explore to harness the role of resources in the circular economy with practices to reduce, reuse, recycle and renew waste. First, **reduce** by increasing material use efficiency that limits waste. This is based on the idea that in a circular economy, using fewer materials results in less waste. This is made possible by the advancements in additive manufacturing, a unique process that involves adding layers to create precise geometric shapes instead of traditional methods that involve removing materials through carving, chipping, or milling. Additive manufacturing technologies, such as 3D printing and computer-aided design, are gaining popularity and are becoming integral to the manufacturing infrastructure. This enables innovative product developments worldwide that embrace the additive manufacturing process. For example, Norsk Titanium has pioneered titanium production methods that reduce waste by half. Companies like Hy5 Pro and Fieldmade leverage 3D printing to minimise material usage when producing prosthetics and spare parts for the energy industry, respectively. Bright

Products focuses on designing multifunctional solar lamps that use fewer raw materials, while Nordic Comfort Products manufactures chairs from recycled plastic waste.

Second, **reuse** through smarter distribution systems to optimise use. A prime example of reuse can be seen in libraries, where hard copy books are read by multiple individuals throughout their product lifespan, providing value to readers. Expanding on this reuse practice, we can explore entrepreneurial avenues. Nabohjelp, for instance, facilitates borrowing instead of purchasing home maintenance tools and equipment, allowing individuals to make use of these items without the need for ownership. Fjong offers a platform for renting clothes, enabling customers to access fashion items without having to buy them outright. The Tise app facilitates gifting and reselling of used clothing, encouraging the reuse of garments. Hayk motivates electric vehicle owners to share cars, reducing the necessity for private car ownership within the same neighbourhoods. In Singapore, there is also a car sharing concept (such as Blue SG and Smove) that emerged in the 1990s, which aligns with the growing ride-sharing economy and contributes to reshaping urban mobility. These initiatives not only promote sustainability but also encourage walking and cycling by providing alternative options for transportation.[22]

Third, **recycling** to produce new products from used materials. Together with customers' involvement in contributing the used bottles and cans, TOMRA Collection Solutions, Norway recycles close to 97% of those in circulation. Norsk Gjenvinning processes recycled paper and the quality is no less inferior as the virgin paper. Norsk Hydro uses 5% of original energy by using energy-efficient processes and recycled aluminium fuelled by renewable energy. Unfortunately, Singapore recycling rate is at a grim 13% for the period of 2020–2021 even as households generated more waste during COVID-19. Compared to previous years, the recycling rate had reached a high of 22% while the target is to achieve 30% recycling rate by 2030. One of the most implacable challenges is the contamination of recyclables in its blue bins. These bins have been around since 2014 but contamination rate is as high as 40% of the gathered waste that becomes non-recyclable. Awareness of recycling is necessary including households separating and disposing their trash, in these blue bins at the foot of every housing block. In the face of repeated

[22] Liew (2022).

reminders from the authorities, executing other options, such as manned recycling booths, together with education outreach campaigns to increase awareness.[23] Embracing a multi-pronged style might be a better way to clean up the mess. Mediacorp, which is Singapore's mainstream broadcasting channel is committed to sustainability and the CNA Green Plan is yet another indication of that commitment. An Online Town Challenge where individuals and neighbourhoods include their green action entries via CNA Facebook, Instagram and TikTok while spreading more awareness at the ground are some of the on-going activities that are launched all year round.[24]

Lastly, renew by turning waste into resources. To achieve renewal, it is essential to transform waste into valuable resources rather than allowing it to contribute to pollution and waste streams. Every industry generates waste, and by extracting valuable resources from this waste, we can minimise our reliance on natural resources. Several innovative examples demonstrate how commercial companies are actively engaged in waste renewal. Finnfjord, for instance, utilises carbon dioxide emissions from its ferrosilicon production to cultivate algae, effectively turning a waste product into a valuable resource. Hias How2O employs a biofilm system that harnesses naturally occurring bacteria to treat wastewater, eliminating the need for hazardous chemicals and effectively removing nitrogen, phosphorus and organic substances from the water. Marealis collects discarded prawn shells from the seafood industry and transforms them into clinically proven natural supplements that aid in lowering blood pressure.[25] Singapore can take further steps to incentivize the adoption of novel circular economy technologies and learn from other economies that have made progress in this area. By doing so, Singapore can strive towards becoming a fully circular economy, aligning with the goals of industry, innovation and infrastructure development (SDG 9). Encouraging the transfer of circular economy technologies and fostering innovation will be crucial in driving Singapore's transition towards a more sustainable and resource-efficient future.

It is evident that Singapore is putting forth significant efforts to enhance its green initiatives and sustainability practices. NParks continue

[23] SPH Media Limited. (2022, April 24).

[24] Mediacorp Pte Ltd (2022).

[25] The Explorer (2020, March 12).

to spearhead the projects of today including promoting walking and cycling. Several city visions (*Garden city* in 1963, then, *City in a Garden* City, *City in Nature* and *Singapore Green Plan 2030* to brand Singapore a *liveable city*[26]) have been launched so as achieve sustainable urban development and subdue the getting larger climate change concerns. Singapore's efforts to restore the natural environment demonstrate its commitment to sustainable development and achieving the SDGs. Advanced economies including Singapore are constantly looking for ways to upgrade their capabilities that transform its past degeneration into regenerative processes.

Nation building efforts had helped Singapore to become economically resilient, but less attention was placed in addressing the degradation of social and natural systems, and negative ecological footprints. Despite so, there are places of opportunity enabling Singapore to overcome some of these challenges. It seems that seismic changes to change the city mindset, design and its built environment is underway, though it does not have to be always actioned by public agencies, as long they can convince individual residents to partake in the nation's green plan journey. Of course, a different tactic to urbanism can also be applied by nurturing transformative changes to address the planetary problems, and indirectly encouraging walking and cycling among the Singapore community.

The evolution of design methodologies for the built environment, considering both social and ecological aspects, has yielded promising outcomes. In addition to cars, walking and cycling have emerged as viable transportation options. The concept of regenerative sustainable built environments revolves around creating urban spaces that not only contribute positively to healing and enhancing overall urban conditions but also allow human activities to coexist harmoniously with natural systems, resulting in mutual benefits. Regenerative sustainable urbanism is a multidisciplinary field that demands a different mindset and approach, incorporating comprehensive, integrative and holistic design strategies to address challenges such as climate change and rapid urbanisation. It presents an opportunity and a pressing need for designers and practitioners to create livable, healthy and sustainable cities that not only mitigate negative ecological impacts but also actively regenerate social and ecological systems.

[26] A+U (2012); Civil Service College Singapore (2008, June 1).

Embracing a circular economy approach presents Singapore with opportunities to explore innovative designs, business models and supply chain strategies within its urban environment. The country has been actively pursuing waste reduction, reuse, recycling and renewal initiatives. Notable breakthroughs have been achieved in the areas of reuse and recycling, although the pandemic has impacted recycling efforts. To foster innovation and rejuvenate business models that are well-suited for Singapore's limited land resources, it is crucial to foster partnerships and collaborations for the transfer and enhancement of circular economy technologies.

References

A+U Special Edition. (2012). *Interview: Poon Hong Yuen, CEO, National Parks Board.*

Auger, T. (2013). *Living in a garden: The greening of Singapore.* National Parks Board.

Baccini, P., & Brunner, P. H. (2012). *Metabolism of the anthroposphere: Analysis, evaluation, design.* MIT Press.

Beatley, T., & Wheeler, S. M. (Eds.). (2004). *The sustainable urban development reader.* Routledge.

Broto, V. C., Allen, A., & Rapoport, E. (2012). Interdisciplinary perspectives on urban metabolism. *Journal of Industrial Ecology, 16*(6), 851–861.

Chan, L., Davison, G., & Chia, C. (2010). *Biodiversity conservation in Singapore. City Green*, Issue 1. National Parks Board.

Chew, V. (2016). Singapore green plan. *Singap. Infopedia.*

Civil Service College Singapore (2008, June 1). *A city in a garden.* https://www.csc.gov.sg/articles/a-city-in-a-garden. Accessed on 26 December 2022.

De Jong, M., Joss, S., Schraven, D., Zhan, C., & Weijnen, M. (2015). Sustainable–smart–resilient–low carbon–eco–knowledge cities; making sense of a multitude of concepts promoting sustainable urbanization. *Journal of Cleaner Production, 109*, 25–38.

DesignSingapore Council (n.d.). *Singapore: City of design, by design—Green spaces matter.* https://designsingapore.press.camronpr.com/singapore-city-of-design-by-design-green-spaces-matter. Accessed on 26 December 2022.

Gosling, E., & Williams, K. (2010). Connectedness to nature, place attachment and conservation behaviour: Testing connectedness theory among farmers. *Journal of Environmental Psychology, 30*(3), 298–304.

Housing and Development Board. (2013). *Landscape guide.* Development and Procurement Group.

Liew, C. (2022). Will car-sharing replace car ownership in Singapore?. https://vibrantdot.co/will-car-sharing-replace-car-ownership-in-singapore/. Accessed on 21 May 2022.

Ma, L., & Guek, L. (2012). *A delightful eco-nature neighbourhood park*. CityGreen #6, Centre for Urban Greenery and Ecology Publication, CUGE.

Mediacorp Pte Ltd. (2022). *About CNA Green Plan*. https://www.channelnewsasia.com/cna-green-plan. Accessed on 21 May 2022.

Newman, P. (2014). Biophilic urbanism: A case study on Singapore. *Australian Planner, 51*(1), 47–65.

Newton, P. W. (2007). Horizon 3 planning: Meshing liveability with sustainability. *Environment and Planning B: Planning and Design, 34*(4), 571–575.

Newton, P., Newman, P., Glackin, S., & Trubka, R. (2012). Greening the greyfields. *International Journal of Social, Behavioral, Educational, Economic, Business and Industrial Engineering, 6*(11), 2870–2889.

NParks Rail Corridor. (2023, April 3). *Our Rail corridor*. https://railcorridor.nparks.gov.sg/. Accessed on 8 April 2023.

Open Government Products. (2022, April 19). *Circular economy*. Circular Economy (towardszerowaste.gov.sg). Accessed on 15 May 2022.

PUB. (2022). *Active, beautiful, clean waters programme*. https://www.pub.gov.sg/abcwaters/about. Accessed on 18 May 2022.

SPH Media Limited. (2022, April 24). *Recycling efforts shouldn't go to waste*. https://www.straitstimes.com/opinion/st-editorial/recycling-efforts-shouldnt-go-to-waste. Accessed on 21 May 2022.

The Explorer. (2020, March 12). *How the circular economy is changing business*. https://www.theexplorer.no/stories/renewable-resources/an-introduction-to-the-circular-economy/. Accessed on 17 May 2022.

United Nations-Habitat. (2013). *State of the World's Cities 2012–2013 report*. https://unhabitat.org/sites/default/files/download-manager-files/State%20of%20the%20World%20Cities%2020122013.pdf. Accessed on 19 November 2022.

Urban Redevelopment Authority. (2022, May 18). *Master Plan*. https://www.ura.gov.sg/Corporate/Planning/Master-Plan. Accessed on 18 May 2022.

Yok, T. P. (2012). *Vertical garden city, Singapore*. https://www.greenroofs.com/2013/11/27/vertical-garden-city-singapore-by-tan-puay-yok/. Accessed on 16 May 2022.

CHAPTER 13

Knowledge on the Way to Walkability

This book delves into the topic of walkability in the built environment of Singapore. It explores how the state has dealt with the challenges and trade-offs in shaping a walkable city. One of the most persistent urban problems in Singapore, as well as other large cities worldwide, is the issue of space and the management of spatial arrangement to accommodate housing, industrial, retail and social needs. In the case of Singapore, the government has played a significant role in shaping Singapore's built environment. But the process is not one-dimensional as the state continuously involves negotiation and consultation with stakeholders within the society. Singapore recognises that urban development cannot be achieved solely through political and economic means. It also involves promoting a range of ideologies that are perceived as beneficial and appropriate, which requires the government to use its power to construct and defend proposals and plans that facilitate the realisation of economic, social and political goals. The built environment has to be functional and cater to the interests of various social and ethnic groups within the city, allowing them to achieve their socioeconomic aspirations. Inclusivity is a crucial consideration in the development and redevelopment of the urban landscape, ensuring that it accommodates and represents the diverse needs and perspectives of the city's residents.

Melissa Liow Li Sa and Sam Choon-Yin, *Sustainable Urban Development in Singapore*, Sustainable Development Goals Series,
https://doi.org/10.1007/978-981-99-5451-3_13

Creating a city that meets the needs of its inhabitants is a complicated process that requires striking a balance between competing interests, effective leadership and stakeholder engagement. Moreover, it is necessary to establish the institutional capacity to ensure that plans and projects are executed successfully. In the words of Cheong Koon Hean, former CEO of HDB and URA, the idea is not to create a "planned city" according to a preconceived blueprint. It is about "a city that plans continually", "at times taking the necessary sharp turns in response to changing situations, but always having the long-term vision and end goal in mind".[1]

It is anticipated that by 2030, the majority of the world's population, roughly 60%, will be living in cities.[2] This underscores the importance of implementing sustainable urban planning and policies that ensure equitable access to infrastructure and social services. It is important to note that every city has its own distinct features and what works in one city may not be successful in another. Nevertheless, we can still draw upon successful practices that can be adapted to various settings.

Singapore's Model

To understand how cities work, a common approach is to study the smaller parts, for example, into functional areas like transport, economic, ageing society, based on the assumption that when the smaller parts are assembled, the whole will approximate the sum of its part. As Peter Ho, former Head of the Singapore Civil Service, points out, in complex systems like cities, the whole is often quite distinct to the components that are part of it thereby making it difficult to predict the outcomes of the whole based on the results from the properties of the components.[3] The reason, Ho explains, is that everything is connected and related to each other, referring to the thoughts of Lao Tze, the ancient Chinese philosopher. "Agents—people—are not independent. Instead, they are interdependent. They interact and influence on another in ways that defy a deterministic or linear analysis. Their interactions lead to outcomes that are inherently unpredictable ex ante, and that are revealed only when they

[1] Cheong (2019, p. 29).

[2] Up from 50% when the report by Ernst & Young was published in 2013. See Ernst & Young (2013).

[3] Ho (2021, October 4).

occur. In other worlds, we know what is going to happen only when it happens. This is the property of emergence, which characterises complex systems". The challenge to urban planners is to look at the complex system not just in their component parts, but also as a whole.

Walkability can be seen as a complex system due to its various components and the interconnected nature of those components. It involves not only the design of streets and sidewalks, but also factors such as land-use patterns, transportation infrastructure and social and cultural aspects. These factors can interact in unpredictable ways, making it difficult to predict the outcomes of changes in one area. Additionally, people's behaviour and preferences can influence walkability, as they may choose to walk or not based on factors such as safety, comfort and convenience. Thus, improving walkability requires a holistic approach that takes into account the various components of the system and their interactions, rather than a simple linear or deterministic analysis.

Singapore recognises the challenge and has begun to plan holistically, taking advantage of high-performance computing, data analytics and artificial intelligence. The urban planning process considers economic, social and developmental perspectives, using scientific tools such as agent-based modelling with the help of data, which is supported by Singapore's Smart Nation initiative. The initiative captures large amount of data via the network of interconnected sensors and processed by high performing computing and powerful data analytics to provide useful information to review existing patterns and discover new ones. Technology advancements had enabled the development of new science of cities.[4] Singapore is at the forefront in improving urban design and services through digital transformation.

[4] The ignorance of science has been a great concern. Jane Jacobs (2005, p. 129) wrote about the case of the boulevard, which can serve "a full range of mobility: walks for pedestrians, lanes alongside them for bicycles and roller skaters; lanes for public transit vehicles, and separate channels for automotive vehicles passing through and those heading for local destinations". Boulevards offer a venue for people to meet each other, enjoy beverages and chat "while they take in the passing scene, including sidewalk play for children" (ibid., p. 130). But traffic engineers in the United States outlaw boulevard trees and other features on safety ground when scientific evidence suggests otherwise; those boulevards the world over are safe for all kinds of mobility. Jacobs attributed this to urban designers' disconnection with science, and the inability to accept alternative views which scientists offer because they contradict with their point of view and might lead to change in paradigm.

One of the key initiatives to promote walkability in Singapore is the Walk2Ride programme, an initiative under the Land Transport Master Plan 2013, which aims to encourage more people to use public transport by making walking to and from public transport nodes more convenient and attractive. This programme involves creating covered walkways, installing pedestrian crossings and providing amenities such as benches and bicycle parking facilities. Another initiative is the "Green Man+" scheme, which uses technology to give elderly pedestrians and persons with a disability up to 13 more seconds to crossroads safely, especially at busy intersections. The government has also been investing in the creation of park connectors and green spaces, which not only encourage walking and cycling but also improve the overall liveability of the city. Furthermore, the government has designated certain areas in the city as "car-lite" zones, where cars are restricted, and pedestrian-friendly amenities are prioritised. One example is the Marina Bay area, which has been transformed into a vibrant pedestrian hub with wide sidewalks, public art installations and green spaces.

To ensure that the needs and preferences of all residents are taken into account, the government has been conducting extensive community engagement and consultation sessions, for example, the 'Our Singapore Conversation' survey conducted by the Institute of Policy Studies and HDB's Sample Household Surveys on residents' satisfaction levels with housing. This enables residents to voice their opinions and concerns about the proposed changes and ensures that their needs are considered in the planning and implementation process. The government works closely with various stakeholders, including town councils, architects and urban planners, to ensure that the designs and layouts of public spaces are conducive to walking and cycling. This involves careful consideration of factors such as the placement of pedestrian crossings, the provision of shade and shelter and the integration of public art and greenery.

Singapore's approach to walkability is a holistic one that involves a range of strategies and initiatives and participation and input of various stakeholders. By taking a multi-pronged and collaborative approach to urban planning, the city-state is able to create more liveable and sustainable environments that promote the health and well-being of its residents.

The Role of the Government

The case of Singapore exemplifies urban planning under inclusive state capitalism. Through regulations and workings of the statutory boards such as LTA, URA and HDB, Singapore is able to create specular urbanscapes to meet the demands of the economy and its people, including the preference for walkability sites and infrastructure. Singapore's model of urban planning under state capitalism is aptly summarised by Gavin Shatkin: "subjugation and co-optation of civil society to the interests of the state; the severe curtailment of alternative claims to urban space outside of the state and the corporate economy; the imposition of a comprehensive regime of state social control; the assertion of state hegemony in determining the aesthetic and functional form of the city; and the bending of the meaning of history and culture to the interests of the state".[5] Public servants in Singapore are expected to comprehend the priorities and agenda of the elected government, despite not being directly involved in the weighing of political considerations.

Single-party rule like the case of Singapore can be useful for seeing through long-term plans because it provides a stable and consistent political environment where policies and plans can be implemented without interference from opposition parties. In a multi-party system, there can be frequent changes in government and shifts in policy priorities, which can make it difficult to implement long-term plans that require sustained effort and resources. Single-party rule can also provide greater control and coordination over government agencies and departments, allowing for more efficient implementation of policies and plans.

It is important to note that single-party rule can also lead to a lack of accountability and transparency, as there may be limited checks and balances on the ruling party's power. This can result in the potential for corruption and abuse of power, which can have negative consequences for the implementation of long-term plans. Singapore has so far avoided the negative consequences of single-party rule through several measures. Firstly, Singapore has a strong legal and institutional framework that promotes transparency and accountability. The Corrupt Practices Investigation Bureau (CPIB), an independent agency, investigates and prosecutes corruption cases, including those involving government officials. The CPIB operates without fear or favour and reports directly to

[5] Shatkin (2014, p. 135).

the elected President. Secondly, Singapore's ruling party, the PAP, has consistently demonstrated a commitment to good governance and anti-corruption measures. The PAP has implemented measures to prevent corruption, such as paying competitive salaries to public officials to reduce the potential for bribes. The party has also taken a strong stance against corrupt officials, even if they are members of the party.[6]

Singapore's widely heralded achievements in urban planning are a consequence of the state's ability and willingness to adapt to changing economic and social environments. Take car population control as an example. Unlike major cities in Southeast Asia, Singapore imposes deliberate policies and regulations to control the growth of car population. In 1975, the Area Licensing Scheme (ALS) was introduced to require cars to pay to a certain fee for entering a defined and restricted zone of the CBD. In 1998, the ALS was replaced by the Electronic Road Pricing Scheme, which imposed a fee on cars for assessing roads, expressways leading to the CBD during peak hours. In 1990, the Motor Vehicle Quota System (VQS) was introduced to control the growth of car population in Singapore. Car buyers are required to obtain a Certificate of Entitlement (COE) valid for up to 10 years.[7] In another case, while the suggestion of having 6.9 million people living in Singapore as reported in the Population White Paper in January 2013 has generated widespread public

[6] The main pillar of Singapore success lies in the country's sterling leadership efforts to ensure the country's stability, future viability and competitiveness. Stemming from the attention to the smallest details and actions carried out in an authoritarian way, the country's first prime minister, Mr. Lee Kuan Yew's and the old guards implemented very effective policies that enabled Singapore to be lifted from the third world to the first, raising Singapore's profile and its national brand accordingly. Lee Kuan Yew has spoken of the importance of excellence in public service as integral. Singapore recognises that the single most important factor to rapid development is the quality of the people in the public service. A bright person who works with the constituents and is capable of gathering a group of other capable persons makes a difference between success and failure of a project. That the government officials act on the feedback from the residents is the hallmark of the Singapore government.

[7] Singapore must keep its little city living and breathing, Kishore Mahbubani wrote. "This is why from early in our history, our founding fathers recognised that our little city-state could choke and die if we allowed too many cars on Singapore roads. If Singapore were full of traffic jams like Bangkok or Jakarta, its economy would grind to a natural halt. We would die just like any human being who tried to swallow too much". Transportation and land-use planning to control car population are supplemented by the provision of mass rapid transit and extensive public transportation networks. See Mahbubani (2017, p. 311).

concerns about the potential rise in prices and the insufficient infrastructure to accommodate the increased population, Singapore recognises that long-term planning is necessary to ensure that infrastructure development and urban renewal keep up with population growth and provide sustainable green spaces and nature reserves. It is not an impossible task, as there are megacities with more than 10 million people, such as Tokyo and Jakarta which has a physical size similar to Singapore and a sizeable spread of public parks and gardens. The government has acknowledged the concerns raised by the public about the pace of population growth and immigration and has committed to continuing the conversation on this matter.

In the past, HDB housing estates contain a large area of space for roads for motorised vehicles, rejecting the use of space for recreation, social interaction and activity mobility users. Through persistent concerns and the fact that housing shortage problem is largely resolved, increasingly more space is devoted to pedestrians and cyclists. Park connectors are extensively developed around the island. In addition to the provision of public facilities like playgrounds and lakes to promote social integration, public housing estates in newer towns are served by more comprehensive cycling and pedestrian networks. Singapore has moved forward by creating car-free housing estates and new towns, for example in Tengah, where free space at the ground level is used for retail and recreational purposes, separating from the roads and traffic which are confined in spaces beneath the town centre. Tengah town centre, wrote Chua Mui Hoong, "will be a showcase of sustainable mobility, putting Singapore into the ranks of future-oriented cities that are designing such car-free town centres".[8] The Jurong Lake District is envisioned as Singapore's second central business district, with an extensive underground network of freight tunnels designed to facilitate the smooth flow of goods to homes and workplaces. The idea is that by having warehouses located underground, the cost and environmental impact of delivery will be reduced. This approach is based on the assumption that online shopping

[8] Chua (2022, June 3). Interestingly, Chua links urban renewal in favour of walkability to winning votes in election. Citing the case of Paris, Chua wrote, "Paris Mayor Anne Hidalgo was elected in 2014. That year, air pollution was so bad in Paris that media headlines compared it to Beijing. Mayor Hildalgo made Paris greener by converting roads into pedestrianized areas and cycling paths and won a second term in 2020".

will become the dominant mode of retail, making physical stores unnecessary. This trend has been further reinforced by the COVID-19 pandemic, which has seen a surge in online grocery shopping.

In 2019, LTA unveiled the Land Transport Master Plan (LTMP) 2040, which outlines the future policies and objectives for the country's transportation system through 2040 and beyond. The plan aims to create a transport network that is fast, well connected and convenient, while also fostering a culture of gracious behaviour and inclusivity in the infrastructure. There has been significant progress in making the environment more accessible to people with different abilities by emphasising walkability. Projects that have made public roads, taxi and bus shelters and walkways more accessible to individuals with disabilities have been successfully completed. Larger and clearer signage is being installed to promote walking and to facilitate safe and convenient travel for vulnerable pedestrians such as the elderly and children. In addition, the government introduced a public transport concession card for people with disabilities, and all new public buses now come equipped with digital Passenger Information Display Systems that assist individuals with hearing and vision impairments in their travels, culminating in a more effortless travel experience.

To promote mutual respect among users, various publicity and educational campaigns are implemented. One such initiative is the Singapore Road Safety Month 2021 campaign launched on 28 May 2021 by the Singapore Road Safety Council (SRSC) and Traffic Police (TP), with support from various organisations like the People's Association, Ministry of Education and LTA. The campaign's primary objective was to raise awareness about speeding offences and red light running, which aligned with the campaign's theme, "Road Safety for All". As part of the campaign, a COVID-19 care pack containing road safety tips is distributed to motorists who perform acts of kindness on the road or exhibit safe driving behaviour to allow pedestrians and cyclists to move around safely and comfortably.

There are communal spaces for individuals to socialise and bond after work and outside their homes. These spaces include inter-generational playgrounds and exercise areas, coffee shops, schools and supermarkets located near residential blocks. Engaging in gardening has been found to have a positive impact on seniors' mental health, and this is reflected in the prevalence of community gardens, such as the Community-in-Bloom projects, in HDB neighbourhoods. Building on this concept, NParks and the National University Hospital's Department of Psychological Medicine collaborated to design a therapeutic garden at HortPark, which offers horticultural therapy to aid in the recovery of post-stroke patients and dementia sufferers, as well as activity spaces for planned programmes.

The state of the environment has become a progressively significant concern for the Singapore government as the country continues to undergo transformation. NParks was tasked to implement the city visions (*Garden city* in 1963 and at present, *City* in a Garden to brand Singapore a liveable city) and take charge of the environmental development. The City in a Garden vision is a comprehensive approach to mitigating the effects of climate change and promoting urban sustainability in Singapore. One of the ways this vision is being realised is through the reconstruction of streetscapes and the restoration of urban parks. Parks are not just green spaces, but they also serve a variety of social and environmental functions, such as enriching biodiversity, serving as carbon sinks and testing sustainable features to address tropical climate challenges like fast evaporation, high temperatures and heavy rainfall. Hill parks are equipped with special amenities like the Southern Ridges, which connect hill parks with two elevated bridges, boardwalks and trails. Visitors can enjoy observing the flora and fauna in their natural habitats, promoting the protection and appreciation of these nature sanctuaries. Another example is the eco-nature neighbourhood park, such as the Greenwood Sanctuary at Admiralty, which is designed with greeneries to recover from microclimatic conditions, drainage and land forming systems and acts as an added green lung for compact built environments.[9] Walkability is key to fulfilling Singapore's biophilic vision and SDGs.

The City in a Garden vision includes an impressive initiative known as the Active, Beautiful and Clean (ABC) Waters programmes, which was launched by PUB in 2006. The programmes aim to unlock the hidden potential of Singapore's reservoirs and waterways, integrating water infrastructure into the urban landscape. This transformation is intended to serve as the core water channel for communities while allowing those in contact with its waterways to enjoy it amid the city's surroundings. The Marine Barrage is one such water infrastructure that serves as flood control, a water reservoir and offers various recreational activities for the community.

Enhancing walkability is often considered a key measure of urban resilience and liveability. A walkable city offers an environment that is conducive to physical activity, including opportunities for social interactions, which as Richard Florida has argued, is crucial to attract professional talents in the creative industries. Singapore's success as a global business hub can be attributed to its ability to attract and retain a diverse pool of talent from around the world. Professionals in the creative industry tend to gravitate towards places that facilitate communication, exchange

[9] In 2008, Singapore launched the Leisure Plan to promote a healthy and active lifestyle for its citizens through the development of various leisure and recreational facilities and activities. The plan aims to provide Singaporeans with a wide range of options for leisure activities, and to promote an active and healthy lifestyle, for example, through the development of the park-connector network, connecting green spaces with walking pathways and cycling routes. The plan encourages citizens to explore the nature and natural heritage of Singapore, to experience the country's rich biodiversity and scenic landscapes. In addition to parks, gardens and heritage, the plan consists of other components, which include:

- Sports: The government promotes sports and physical activity through the development of sports facilities, such as stadiums and swimming pools, as well as through the organisation of sports events and competitions.
- Arts and Culture: The government supports the arts and culture scene in Singapore through the development of museums, art galleries and performing arts centres.
- Community Centres: The government has established community centres across the island to provide citizens with access to a wide range of leisure activities, such as fitness classes, sports and cultural activities.
- Entertainment: The government supports the development of the entertainment industry in Singapore through the promotion of a diverse range of entertainment options, such as theme parks, movie theatres and nightlife venues.

The Leisure plan is an integral part of Singapore's effort to create a liveable city for its citizens, enhancing the quality of life and promoting a sense of community.

of ideas, collaboration and networking, as they value the exchange of ideas and the formation of professional relationships. Singapore's vibrant cultural scene, talented workforce and beautiful natural surroundings provide ample opportunities for external stimulation and engagement, while the availability of peaceful and serene spaces allows for solitude and reflection. The provision of communal spaces in the city such as coffee shops, co-working spaces and public parks aligns with Singapore's vision to become a thriving hub for the creative class. Notably, choosing a place to live is an important decision that can significantly impact various aspects of our lives. It can influence our income across subsequent life stages, social circles and opportunities for our families, among other things. In this regard, Singapore has made impressive strides in developing a built environment that is not only functional and aesthetically pleasing, but also conducive to walkability.

Indeed, Singapore is home to people from different ethnic and cultural backgrounds who live and work side-by-side in harmony. This diversity is reflected in the city's built environment, which features a wonderful mix of natural features, heritage buildings and modern architecture. This creates a vibrant and dynamic urban landscape that is both visually appealing and culturally rich. Indeed, a fond memory of Singapore is likely to be its superb cuisine. Singapore is renowned for its street food, which is sold at hawker centres and food courts throughout the city and within walking distance in most cases. These food centres offer a wide variety of dishes from different cultures and at affordable prices. Walking around Singapore, one can easily find a place to satisfy their taste buds without having to travel far.

Singapore is blessed with abundant green space, which makes it a pleasure to walk around the city. It is also endowed with a tropical climate all year round with beautiful shorelines, and spared the ravages of nature such as earthquake, typhoon and volcanic eruption abundance greenery. The city-state has made significant investments in developing magnificent urban parks, trails and islands, which provide residents with ample opportunities to enjoy the great outdoors. These parks are designed to be inclusive where cyclists, strollers and picnickers can congregate together. This creates a sense of community and belonging, which is an essential aspect of walkability. They are well-maintained and clean, which makes them pleasant places to spend time in. While the weather can be challenging at times, Singapore has developed a comprehensive network of pedestrian-friendly infrastructure, such as shaded walkways and underground pathways that allow people to move around the city with ease and comfort. Overall, Singapore has done well in the area of environmental sustainability. Foo Tuan Seik wrote of Singapore's "high quality urban environment" attributed particularly to its ability to control pollution.[10] Soh Suat Hong, notes that as early as "the 1980s, basic infrastructure to meet Singapore's environmental needs, such as removal of solid waste, waste water and storm water, were in place".[11]

Singapore's pedestrian-friendly space and infrastructure are also noteworthy. The city-state has developed an extensive network of sidewalks, crosswalks and pedestrian bridges that allow people to move around the city safely and comfortably. The city also has a comprehensive public transportation system that includes buses, trains and taxis, which provides residents with convenient and affordable options for getting around. Singapore's low crime rate makes it a safe place to live and work.

To conclude, it is worth reminding ourselves that environmental concerns are on the rise. It is often assumed that individuals lack the power to effect change in times of moral crisis, as they are so dependent on collective action. However, the COVID-19 pandemic has shown that individuals can make a significant difference, as seen in the power of wearing masks. If everyone plans strategically and plays their part, no matter how small, the world can be a safer and better place. It is important to acknowledge that bad trends often contain the seeds of reversal, as the growing social costs they impose can lead to a greater willingness to seek curative measures. Each of us has a crucial role to play, no matter how small it may seem. Singapore is a prime example of how environmental degradation can spur various stakeholders within the community the search for alternatives to car usage and the advancement of walkability. It may be beneficial to study the Singaporean case in this regard.

References

Cheong, K. H. (2019). *Seeking a better urban future*. World Scientific Publishing Company.

Chua, M. H. (2022, June 3). Let's talk about inequality in mobility options. *The Straits Times*.

[10] Foo (1996).

[11] Soh (2000, p. 235).

Ernst & Young. (2013). *Hitting the sweet spot: The growth of the middle class in emerging markets*. https://www.wall-street.ro/files/148266-435.pdf

Foo, T. S. (1996). Urban environment policy: The use of regulatory and economic instruments in Singapore. *Habitat International, 20*(1), 5–22.

Ho. P. (2021, October 4). Singapore and the Science of Cities. *The Straits Times*.

Jacobs, J. (2005). *Dark age ahead*. Vintage Books.

Mahbubani, K. (2017). Singapore: the smartest city on our planet. In Heng, C. K. (Ed.) *50 years of urban planning in Singapore*. World Scientific Publishing Corporation.

Shatkin, G. (2014). Reinterpreting the meaning of the 'Singapore model': State capitalism and urban planning. *International Journal of Urban and Regional Research, 38*(1), 116–137.

Soh, S. H. (2000). *Singapore. In Asian approach to resource conservation and environmental protection*. Asian Productivity Organization.

Exhibit

Cycling path specifically designated along Bukit Panjang Ring Road

Melissa Liow Li Sa and Sam Choon-Yin, *Sustainable Urban Development in Singapore*, Sustainable Development Goals Series,
https://doi.org/10.1007/978-981-99-5451-3

Cycling parking area at Hougang Bus Interchange

Designated bicycle parking area within a public housing estate

Indoor Bus Interchange with Air Conditioning

Hawker culture at Gluttons Bay

bak kut teh (Pork Bone Tea soup)

Residents enjoying a leisurely walk in the Singapore Botanic Gardens

The Rail Corridor near The Rail Mall along Upper Bukit Timah Road, where walkability and scenic beauty converge

Wide pedestrian pathways at Singapore's Orchard Road Shopping Strip

One North's Biopolis featuring abundant green spaces and pedestrian-friendly spaces

Chinatown in the Day and Night

Tourists flocking the Merlion at Marina Bay

Tourists and residents walking along the Marina Bay area

Designated Silver Zone in Hougang Street 91

Gardens on Rooftop of Marina Square Mall

In the Heartlands—Clementi Town Centre

In the Heartlands—Punggol District and Punggol Park

References

A+U Special Edition. (2012). *Interview: Poon Hong Yuen, CEO, National Parks Board*.

Agency for Integrated Care Pte Ltd. (2022). *What is dementia?* AIC Dementia. Accessed on 8 May 2022.

airssist. (2022). *Top business travel trends 2023*. https://airssist.com/business-travel-trends/. Accessed on 26 December 2022.

Alawadi, K., Striedinger, V. H., Maghelal, P., & Khanal, A. (2021). Assessing walkability in hot arid regions: The case of downtown Abu Dhabi. *Urban Design International*, *27*, 211–231. https://doi.org/10.1057/s41289-021-00150-0. Accessed on 5 January 2023.

Alfred, R., & Goodman, A. (2020). Low traffic neighbourhoods, car use, and active travel: Evidence from the people and places survey of outer London active travel interventions. *Transport Findings*. https://doi.org/10.32866/001c.17128. Accessed on 14 September 2021.

Alonso, T. (2021, January 13). *Success story: The transformation of singapore into a sustainable Garden City*. https://tomorrow.city/a/singapore-transformation-garden-city. Accessed on 17 October 2021.

Ang, Q. (2022, January 23). First half of round Island route opens from seletar to labrador. *The Straits Times*.

Ang, S. (2022, April 4). *Elder abuse on the rise, but no increase in higher-risk cases: Sun Xueling*. https://www.straitstimes.com/singapore/politics/elder-abuse-on-the-rise-but-no-increase-in-higher-risk-cases-sun-xueling. Accessed on 21 March 2023.

Ansoff, H. I. (1957). Strategies for diversification. *Harvard Business Review*, *35*(5), 113–124.

Melissa Liow Li Sa and Sam Choon-Yin, *Sustainable Urban Development in Singapore*, Sustainable Development Goals Series,
https://doi.org/10.1007/978-981-99-5451-3

Arun, M., & Yap, M. T. (2000). Singapore: The development of an intelligent island and social dividends of information technology. *Urban Studies, 37*(10), 1749–1756.

Asian Development Bank (ADB). (2011). *Walkability and pedestrian facilities in Asian cities state and issues*. Asian Development Bank (adb.org). Accessed on 22 August 2021.

Auger, T. (2013). *Living in a garden: The greening of Singapore*. National Parks Board.

Azadi, R., Yousefi, B., & Eydi, H. (2015). The impact of brand country-of-origin image on the formation of brand equity in the sports apparel industry. *Universal Journal of Industrial 3*(3), 67–73. https://doi.org/10.13189/ujibm.2015.030301

Baccini, P., & Brunner, P. H. (2012). *Metabolism of the anthroposphere: Analysis, evaluation, design*. MIT Press.

Ballantyne, R., Packer, J., & Falk, J. (2011). Visitors' learning for environmental sustainability: Testing short-and long-term impacts of wildlife tourism experiences using structural equation modelling. *Tourism Management, 32*(6), 1243–1252.

Balsas, C. J. (2021). Exciting walk-only precincts in Asia, Europe and North-America. *Cities, 112*. https://doi.org/10.1016/j.cities.2021.103129

Baldwin, A., Harris, T., & Davies, G. (2008). Look right! A retrospective study of pedestrian accidents involving overseas visitors to London. *Emergency Medicine Journal, 25*, 843–846.

Barnard, T. P. (2016). *Nature's colony: Empire, nation and environment in the Singapore Botanic Gardens*. NUS Press.

Barter, P. (2013). Singapore's mobility model: Time for an update?. In *institute of mobility research Megacity mobility culture*. Springer-Verlag.

Bartholomew, K., & Ewing, R. (2011). Hedonic price effects of pedestrian-and transit-oriented development. *Journal of Planning Literature, 26*(1), 18–34.

Beatley, T., & Wheeler, S. M. (Eds.). (2004). *The sustainable urban development reader*. Routledge.

Beaverstock, J. V., Derudder, B., Faulconbridge, J., & Witlox, F. (2016). *International business travel in the global economy*. Routledge.

Ben-Joseph, E., Lee, J. S., Cromley, E. K., Laden, F., & Troped, P. J. (2013). Virtual and actual: Relative accuracy of on-site and web-based instruments in auditing the environment for physical activity. *Health and Place, 19*, 138–150.

Benedict, M. A., & McMahon, E. T. (2001). *Green infrastructure: Smart conservation for die 21st Century*. http://www.conservationfund.org/sites/default/files/GI_SC21C.pdf. Accessed on 21 December 2022.

Bensinger, G. (2021, September 29). How digital maps mislead us. *The Straits Times*.

Biewener, A. A., & Patek, S. (2003). *Animal locomotion*. Oxford University Press.

Birch, D. (1994). Cultural studies in the Asia Pacific. *Southeast Asian Journal of Social Science, 22*(1), 1–12.

Blincoe, L., Miller, T. R., Zaloshnja, E., & Lawrence, B. A. (2015). *The economic and societal impact of motor vehicle crashes, 2010* (Revised No. DOT HS 812 013). https://trid.trb.org/view/1311862. Accessed on 2 February 2022.

Bob, G., Hopwood, B., Mellor, M., & O'Brien, G. (2006). Back to the city: A route to urban sustainability. *Future forms and design for sustainable cities* (pp. 13–30). Routledge.

Brookfield, K., & Tilley, S. (2016). Using Virtual Street audits to understand the walkability of older adults' route choices by gender and age. *International Journal of Environmental Research and Public Health, 13*(11), 1–12.

Broto, V. C., Allen, A., & Rapoport, E. (2012). Interdisciplinary perspectives on urban metabolism. *Journal of Industrial Ecology, 16*(6), 851–861.

Brouder, P., & Ioannides, D. (2014). Urban tourism and evolutionary economic geography: Complexity and co-evolution in contested spaces. *Urban Forum, 25*, 419–430.

Budget Direct Insurance. (2023 February). *Car Ownership Singapore 2023.* https://www.budgetdirect.com.sg/car-insurance/research/car-ownership-singapore. Accessed on 25 March 2023.

Bushell, M. A., Poole, B. W., Zegeer, C. V., & Rodriguez, D. A. (2013). *Costs for pedestrian and bicyclist infrastructure improvements*. University of North Carolina Highway Safety Research Center, University of North Carolina, Chapel Hill. http://www.pedbikeinfo.org/cms/downloads/Countermeasure%20Costs Report Nov2013.pdf. Accessed on 8 December 2018.

Calder, K. (2017). *Singapore: Smart city, smart state*. Brookings Institution Press.

Caldwell, N. (2019, May 17). *Urban planners are saving the planet with redesigned, walkable cities*. https://www.greenmatters.com/travel/2018/05/23/22PR7j/urban-planners-cities. Accessed on 1 August 2021.

Calogiuri, G., Litleskare, S., Fagerheim, K. A., Rydgren, T. L., Brambilla, E., & Thurston, M. (2018). Experiencing nature through immersive virtual environments: Environmental perceptions, physical engagement, and affective responses during a simulated nature walk. *Frontiers in Psychology*, 8. https://doi.org/10.3389/fpsyg.2017.02321

Cameron, R. W., Blanuša, T., Taylor, J. E., Salisbury, A., Halstead, A. J., Henricot, B., & Thompson, K. (2012). The domestic garden—Its contribution to urban green infrastructure. *Urban Forestry & Urban Greening, 11*(2), 129–137.

Campbell, W. (1967, August 1). A new life planned for an old fortress. *The Straits Times*, p. 10. https://eresources.nlb.gov.sg/newspapers/Digitised/Article/straitstimes19670801-1.2.71. Accessed on 4 February 2022.

Cannarella, G. (2022, June 8). *Gastro-diplomacy: The science that values food as an identity*. https://www.finedininglovers.com/article/gastro-diplomacy. Accessed on 17 November 2022.

Cantor, A. (2021). Hydrosocial hinterlands: An urban political ecology of Southern California's hydrosocial territory. *Environment and Planning E: Nature and Space*, *4*(2), 451–474. https://doi.org/10.1177/251484862090938

Cappello, N., & Buder, S. (2020, December 30). *No cars allowed: The 10 best pedestrian streets around the world*. https://www.afar.com/magazine/the-10-best-pedestrian-streets-around-the-world. Accessed on 1 August 2021.

Carbone, F. (2019). Post-multicultural challenges for cultural heritage managers and museums in the age of migrations. *Museum Management and Curatorship, 34*(1), 2–23.

Carl H. Lindher College of Business. (n.d.). *What is gastrodiplomacy?* http://www.gastrodiplomacy.net/#home. Accessed on 17 November 2022.

Case, W. (1996). Can the halfway house stand? *Comparative Politics, 28*(4), 437–464.

Cass, N., & Faulconbridge, J. (2017). Satisfying everyday mobility. *Mobilities, 12*(1), 97–115. https://doi.org/10.1080/17450101.2015.1096083

Cavagna, G. A., Heglund, N. C., & Taylor, C. R. (1977). Mechanical work in terrestrial locomotion: Two basic mechanisms for minimizing energy expenditure. *American Journal of Physiology-Regulatory, Integrative and Comparative Physiology, 233*(5), 243–261.

Centre for Liveable Cities and Urban Land Institute. (2014). *Creating healthy places through active mobility*. https://www.clc.gov.sg/research-publications/publications/books/view/creating-healthy-places-through-active-mobility. Accessed on 7 April 2023.

Centre for Liveable Cities and Urban Land Institute. (2017). *Urban mobility: 10 cities leading the way in Asia-Pacific*. https://www.clc.gov.sg/research-publications/publications/books/view/urban-mobility-10-cities-leading-the-way-in-asia-pacific. Accessed on 7 April 2023.

Centre for Liveable Cities. (2017). *Piloting assisted living: Research and workshop insights*.

Cervero, R., & Sullivan, C. (2010). *Toward green TODs*. UC Berkeley Center for Future Urban Transport.

Chan, A. (2021, September 6). *Commentary: How ride-hailing will change to help countries meet climate goals*. https://www.channelnewsasia.com/commentary/grab-transport-sustainability-uber-didi-app-ride-driver-ev-net-zero-2154676#:~:text=While%20ride%2Dhailing%20accounts%20for,US%24230%20billion%20by%202026. Accessed on 28 December 2022.

Chan, L., Davison, G., & Chia, C. (2010). *Biodiversity conservation in Singapore. City Green*, Issue 1. National Parks Board.

Chang, C. Y., Woo, T. H., & Wang, S. F. (2011). Analysis of pedestrian walking speeds at crosswalks in Taiwan. *Journal of the Eastern Asia Society for Transportation Studies, 9*, 1186–1200.

Chang, T. C., & Teo, P. (2001). From rhetoric to reality: cultural heritage and tourism in Singapore. In L. Low & D. Johnstone (Eds.), *Singapore Inc: Public policy options in the third millennium* (pp. 273–303).

Chapple-Sokol, S. (2013). Culinary diplomacy: Breaking bread to win hearts and minds. *The Hague Journal of Diplomacy, 8*(2), 161–183.

Cheng, T. C., Li, J., & Vaithianathan, R. (2019). Monthly spending dynamics of the elderly following a health shock: Evidence from Singapore. *Health Economics, 28*(1), 23–43.

Cheng, X. M. (2017). "City of gastronomy" of UNESCO Creative Cities Network: From international criteria to local practice. *Social Systems Studies*, 55–67.

Cheong, K. H. (2019). *Seeking a better urban future*. World Scientific Publishing Company.

Chua, B. H., (1997). Not depoliticized but ideologically successful: The public housing programme in Singapore. In J. H. Ong, C. K. Tong, & E. S. Tan (Eds.), *Understanding Singapore society*. Times Academic Press.

Chua, J. (2016, January 20). *Here's Singapore's first carbon negative house*. https://www.eco-business.com/news/heres-singapores-first-carbon-negative-house/. Accessed on 4 December 2022.

Chua, M. H. (2022, May 27). How 'jaywalkers' lost the battle for road space to cars. *The Straits Times*.

Chua, M. H. (2022, June 3). Let's talk about inequality in mobility options. *The Straits Times*.

Civil Service College Singapore (2008, June 1). *A city in a garden*. https://www.csc.gov.sg/articles/a-city-in-a-garden. Accessed on 26 December 2022.

Clarke, D. (2016). Theorising the role of cultural products in cultural diplomacy from a cultural studies perspective. *International Journal of Cultural Policy, 22*(2), 147–163.

Clarke, G., & Monk, A. (2010). Investment Corporation (GIC): Insurer of last resort and bulwark of nation-state legitimacy. *The Pacific Review, 23*(4), 429–451.

Clarke, P., Ailshire, R., Melendez, M., Bader, M., & Morenoff, J. (2010). Using Google Earth to conduct a neighbourhood audit: Reliability of a virtual audit instrument. *Heath and Place, 16*(6), 1224–1229.

Chang, C. Y., Woo, T. H., Wang, S. F., & Kuo, M.-J. (2010). *Survey and analysis of walking speeds for the elderly and children at crosswalks*. https://www.researchgate.net/publication/321913430_Survey_and_Analysis_of_Walking_Speeds_for_the_Elderly_and_Children_at_Crosswalks. Accessed on 31 July 21.

Cheam, J. (2009a, December 3). Singapore to pledge 16% cut in gas emissions. *The Straits Times*.

Cheam, J. (2009b, December 3). Singapore wants to help bring about a deal. *The Straits Times*.

Chew, V. (2016). Singapore green plan. *Singap. Infopedia*.

Chiang, H. D. (2015). *Elections in Singapore*. Straits Times Press.

Chong, K. H., Jia, Z., Loo, D., & Cho, M. (2015). Successful aging in high-density city state: A review of Singapore's aging policies and urban initiatives. In F. G. Caro, & K. G. Fitzgerald (Eds.), *International perspectives on age friendly cities*. Routledge.

Cong, L., Wu, B., Morrison, A. M., Shu, H., & Wang, M. (2014). Analysis of wildlife tourism experiences with endangered species: An exploratory study of encounters with giant pandas in Chengdu, China. *Tourism Management, 40*, 300–310. https://doi.org/10.1016/j.tourman.2013.07.005

Connolly, C., & Muzaini, H. (2021). Urbanizing islands: A critical history of Singapore's offshore islands. *Environment and planning E: Nature and space*. https://journals.sagepub.com/doi/full/10.1177/25148486211051082

Council of Europe. (2022). *Strategy 21—S4 Promote heritage as a meeting place and vehicle for intercultural dialogue, peace and tolerance (Program Description)*. https://www.coe.int/en/web/culture-and-heritage/strategy-21-s4. Accessed on 23 July 2022.

Corlett, R. T. (1995). Rain Forest in the City: Bukit Timah Nature Reserve Singapore. *Gardens' Bulletin Singapore Supplement, 3*, 1–165.

Corlett, R. T. (1997). The vegetation in the nature reserves of Singapore. *Gardens' Bulletin Singapore, 49*(2), 147–159.

Creswell, T. (2010). Towards a politics of mobility. *Environment and Planning D: Society and Space, 28*(1), 17–31. https://doi.org/10.1068/d11407

Cui, C., Liu, Y., Hope, A., & Wang, J. (2018). Review of studies on the public–private partnerships (PPP) for infrastructure projects. *International Journal of Project Management*, *36*(5), 773–794. https://doi.org/10.1016/j.ijproman.2018.03.004

Cusack, J. (2022, September 25). *Singapore financial crime dashboard 2022 by FCN*. https://thefinancialcrimenews.com/singapore-financial-crime-dashboard-by-fcn-2022/. Accessed on 28 January 2023.

Dale, O. L. (1999). *Urban planning in Singapore: The transformation of a city*. Oxford University Press.

Dancel, R. (2021, October 24). Bicycles: A boredom buster and lifeline amid pandemic. *The Straits Times*.

Davies, C., MacFarlane, R., McGloin, C., & Roe, M. (2006). Green infrastructure planning guide. *Project: Final Report*. https://www.researchgate.net/profile/Maggie-Roe/publication/265012095_GREEN_INFRASTRUCTURE_PLANNING_GUIDE_Authors/links/54f087a70cf24eb87940c8c1/GREEN-INFRASTRUCTURE-PLANNING-GUIDE-Authors.pdf. Accessed on 21 December 2022.

Day, J. (2022). *Overview and impact of Singapore's green plan 2030*. https://www.jonesday.com/en/insights/2022/08/overview-and-impact-of-singapores-green-plan-2030. Accessed on 4 December 2022.

de Brown, L. D., Coteau, D., & Lavrushkina, N. (2020). Taking a walk: The female tourist experience. *Tourist Studies, 20*(3), 354–370.

De Jong, M., Joss, S., Schraven, D., Zhan, C., & Weijnen, M. (2015). Sustainable–smart–resilient–low carbon–eco–knowledge cities; making sense of a multitude of concepts promoting sustainable urbanization. *Journal of Cleaner Production, 109*, 25–38.

Dempsey, K., Gilchrist, R., Ashbee, J., Sagrott, S., & Stones, S. (2020). Beyond the martial façade: Gender, heritage and medieval castles. *International Journal of Heritage Studies, 26*(4), 352–369.

Department of Statistics, Singapore. (2021). *Census of population 2020 statistical release 2: Households, geographic distribution, transport and difficulty in basic activities*. Ministry of Trade and Industry.

DesignSingapore Council (n.d.). *Singapore: City of design, by design—Green spaces matter*. https://designsingapore.press.camronpr.com/singapore-city-of-design-by-design-green-spaces-matter. Accessed on 26 December 2022.

DesignSingapore Council, Ministry of Communications and Information. (2015, October 1). *Design to improve lives of the elderly*. https://www.nas.gov.sg/archivesonline/data/pdfdoc/20151008002/media_release_-_aidi_publications_launch_final.pdf. Accessed on 8 May 2022.

Diaz Soria, I. (2017). Being a tourist as a chosen experience in a proximity destination. *Tourism Geographies, 19*(1), 96–117.

Diaz-Soria, I., & Llurdes-Coit, J. C. (2013). Reflections on proximity tourism as a strategy for local development. *Cuadernos De Turismo, 32*, 65–88.

Division of Population Health, National Center for Chronic Disease Prevention and Health Promotion. (2019, April 5). *What is dementia?* https://www.cdc.gov/aging/dementia/index.html#:~:text=Dementia%20is%20not%20a%20specific,a%20part%20of%20normal%20aging. Accessed on 14 May 2022.

Djabarouti, J. (2020). Listed buildings as socio-material hybrids: Assessing tangible and intangible heritage using social network analysis. *Journal of Heritage Management, 5*(2), 169–190.

Dobbs, S., & Loh, K. S. (2023). The origins of urban renewal in Singapore: A transnational history. *Journal of Urban History, 49*(1), 60–84.

Dolnicar, S., Lazarevski, K., & Yanamandram, V. (2013). Quality of life and tourism: A conceptual framework and novel segmentation base. *Journal of Business Research, 66*(6), 724–729. https://doi.org/10.1016/j.jbusres.2011.09.010

Domrongchai, A. (2022, September 28). *NYC gets its first Hawker Center—With vendors straight from Singapore*. Food & Wine. https://finance.yahoo.com/news/nyc-gets-first-hawker-center-164009557.html. Accessed on 1 January 2023.

Douglas-Jones, R., Hughes, J. J., Jones, S., & Yarrow, T. (2016). Science, value and material decay in the conservation of historic environments. *Journal of Cultural Heritage, 21*, 823–833.

Easton, S., & Ferrari, E. (2015). Children's travel to school—The interaction of individual, neighbourhood and school factors. *Transport Policy*, *44*, 9–18. https://doi.org/10.1016/j.tranpol.2015.05.023

Edensor, T. (2010). Walking in rhythms: Place, regulation, style and the flow of experience. *Visual Studies, 25*(1), 69–79.

Eldeeb, G., Mohamed, M., & Páez, A. (2021). Built for active travel? Investigating the contextual effects of the built environment on transportation mode choice. *Journal of Transport Geography*, *96*. https://doi.org/10.1016/j.jtrangeo.2021.103158

Elkington, J. (2004). Enter the triple bottom line. In A. Henriques & J. Richardson (Eds.), *The Triple bottom line: Does it all add up*. (pp. 1–16.). Earth Scan.

Elvik, R. (2010). Why some road safety problems are more difficult to solve than others. *Accident Analysis and Prevention, 42*(4), 1089–1096.

Erh, J. (2020). *2021/108 COVID-19's Economic impact on tourism in Singapore*. https://www.iseas.edu.sg/articles-commentaries/iseas-perspective/2021-108-covid-19s-economic-impact-on-tourism-in-singapore-by-joey-erh/. Accessed on 9 August 2022.

Ernst & Young. (2013). *Hitting the sweet spot: The growth of the middle class in emerging markets*. https://www.wall-street.ro/files/148266-435.pdf

ExplorersSG. (2020). *Coast-to-coast trail*. https://explorersg.com/listings/coast-to-coast-trail/. Accessed on 23 February 2023.

Faux, J., & Dwyer, L. (2009). *Triple bottom line reporting of tourism organizations to support sustainable development*. . BEST Education Network. University of Technology Sydney.

Fino, M. A., Cecconi, A. C., & Bezzecchi. (2021). *Gastronazionalismo*. People. ISBN 9791280105776.

Fleming, T., Turner, S., & Tarjomi, L. (2013). *Reallocation of road space (No. 530)*. https://trid.trb.org/view/1262392. Accessed on 19 April 2020.

Florida, R. (2002a). Bohemia and economic geography. *Journal of Economic Geography, 2*, 55–71.

Florida, R. (2002b). *The rise of the creative class*. Basic Books.

Foo, T. S. (1996). Urban environment policy: The use of regulatory and economic instruments in Singapore. *Habitat International, 20*(1), 5–22.

Forsyth, A. (2015). What is a walkable place? The walkability debate in urban design. *Urban Design International, 20*(4), 274–292.

Fox, R., & Gee, N. R. (2017). Great expectations: Changing social, spatial and emotional understandings of the companion animal–human relationship. *Social & Cultural Geography, 20*(1), 43–63. https://doi.org/10.1080/14649365.2017.1347954

Franklin, A., & Crang, M. (2001). The trouble with tourism and travel theory. *Tourist Studies, 1*(1), 5–22. https://doi.org/10.1177/146879760100100101

Frumkin, H., Frank, L., & Jackson, R. (2004). *Urban sprawl and public health: Designing, planning, and building for health communities*. Island Press.

Fujii, T., & Ray, R. (2022) Singapore as a sustainability city: Past, present, and the future. In H. T. Hoon (Eds.), *The Singapore economy: Dynamism and inclusion*. Routledge.

Fung, J. C. (2015). *Dementia design sourcebook: Design guide, design elements*. Department of Architecture, National University of Singapore.

Gardens by the Bay. (n.d.). *Energetics of the conservatories*. https://www.gardensbythebay.com.sg/en/about-us/our-gardens-story/sustainability-efforts.html. Accessed on 8 April 2023.

Gaw, L. Y. F., & Richards, D. R. (2021). Development of spontaneous vegetation on reclaimed land in Singapore measured by NDVI. *Plos One, 16*(1). https://www.journals.plos.org/plosone/article?id=10.1371/journal.pone.0245220. Accessed on 3 February 2022.

Gee, C., Arivalgan, Y., & Chao, F. (2018). *Singapore perspectives 2018: Together*. World Scientific Publishing Corporation.

Gehrke, S. R., & Wang, L. (2020). Operationalizing the neighborhood effects of the built environment on travel behavior. *Journal of Transport Geography, 82*. https://doi.org/10.1016/j.jtrangeo.2019.102561

George, C. (2012). *Freedom from the press: Journalism and state power in Singapore*. NUS Press.

Gilderbloom, J. I., Riggs, W. W., & Meares, W. L. (2015). Does walkability matter? An examination of walkability's impact on housing values, foreclosures and crime. *Cities, 42*, 13–24. https://doi.org/10.1016/j.cities.2014.08.001

Goh, D. (2001). The politics of the environment in Singapore? Lessons from a 'strange' case. *Asian Journal of Social Science, 29*(1), 9–34.

Goh, H. C., & Heng, C. K. (2017). Shaping Singapore's cityscape through urban design. In C. K. Heng (Ed.). *50 years of urban planning in Singapore*. World Scientific Publishing Corporation.

Goh, Y. H. (2002, October 18). To thrive, Singapore must stay open while developing own talent: DPM Wong. *The Straits Times*.

Gordin, V., & Trabskaya, J. (2013). The role of gastronomic brands in tourist destination promotion: The case of St. Petersburg. *Place Branding and Public Diplomacy, 9*(3), 189–201.

Gosling, E., & Williams, K. (2010). Connectedness to nature, place attachment and conservation behaviour: Testing connectedness theory among farmers. *Journal of Environmental Psychology, 30*(3), 298–304.

Government of Singapore. (2021a). *Cycling*. https://www.lta.gov.sg/content/ltagov/en/getting_around/active_mobility/walking_cycling_infrastructure/cycling.html. Accessed on 19 September 2021.

Government of Singapore. (2021b). *Joo Chiat*. https://eresources.nlb.gov.sg/infopedia/articles/SIP_946__2008-11-13.html. Accessed on 4 September 2021.

Government of Singapore. (2021c). *LTA annual report 2019/20. Riding it out together*. https://www.lta.gov.sg/content/dam/ltagov/who_we_are/statistics_and_publications/report/pdf/LTA_AR1920.pdf. Accessed on 12 September 2021.

Government of Singapore. (2021d). *Rail Network*. https://www.lta.gov.sg/content/ltagov/en/upcoming_projects.html. Accessed on 19 September 2021.

Government of Singapore. (2021, July 1). *Does the Government have a population target, e.g. 10 million?* https://www.gov.sg/article/does-the-government-have-a-population-target#:~:text=At%20that%20update%2C%20the%20Government,below%206.9%20million%20by%202030. Accessed on 3 February 2022.

Government of Singapore. (2022a). *Clean-up of Singapore River and Kallang Basin*. https://eresources.nlb.gov.sg/infopedia/articles/SIP_2019-05-21_104327.html. Accessed on 18 December 2022.

Government of Singapore. (2022b). *Vehicle ownership policies*. https://www.mot.gov.sg/what-we-do/motoring-road-network-and-infrastructure/vehicle-ownership#:~:text=Launched%20in%20May%201990%2C%20the,through%20an%20online%20open%20auction. Accessed on 25 March 2023.

Government of Singapore. (2022, March 5). *Factsheet: 50 silver zones by 2023 to enhance road safety for senior pedestrians*. https://www.lta.gov.sg/content/ltagov/en/newsroom/2020/3/news-releases/50-silver-zones-by-2023-to-enhance-road-safety-for-senior-pedest.html#:~:text=Together%20with%20the%2017%20Silver,will%20be%20completed%20by%202023. Accessed on 11 May 2022.

Government of Singapore. (2022, April 8). *Electric vehicles*. https://www.lta.gov.sg/content/ltagov/en/industry_innovations/technologies/electric_vehicles.html?gclid=Cj0KCQiAo-yfBhD_ARIsANr56g736-zSBBVL3eDJSad2w4OeZOmHfCJhpLRJ0VHM32issr6OI63nHVIaAlc2EALw_wcB. Accessed on 26 February 2023.

Government of Singapore. (2022, December 8). *Sustainable development goals.* https://www.singstat.gov.sg/find-data/sdg. Accessed on 26 March 2023.

Government of Singapore. (2023a). *Preservation of Monuments Act (CHAPTER 239).* https://sso.agc.gov.sg/Act-Rev/PMA1970/Published/19870330?DocDate=19870330. Accessed on 8 January 2023.

Government of Singapore. (2023b). *Sustainable Development.* https://www.mfa.gov.sg/SINGAPORES-FOREIGN-POLICY/International-Issues/Sustainable-Development . Accessed on 26 February 2023.

Government of Singapore. (2023c). *URA announces it conservation master plan 26th December 1986.* https://eresources.nlb.gov.sg/history/events/c90b40e3-9370-485b-99a6-6f761fed3b35. Accessed on 8 January 2023.

Government of Singapore (2023, January 11). *Certificate of Entitlement (COE).* https://onemotoring.lta.gov.sg/content/onemotoring/home/buying/upfront-vehicle-costs/certificate-of-entitlement--coe-.html#:~:text=Understanding%20COE,-All%20vehicles%20in&text=To%20register%20a%20vehicle%2C%20you%20must%20first%20place%20a%20bid,exercises%20conducted%20twice%20a%20month. Accessed on 26 February 2023.

Government of Singapore. (2023, March 8). *Statistics public transport ridership.* https://www.lta.gov.sg/content/ltagov/en/who_we_are/statistics_and_publications/statistics.html . Accessed on 25 March 2023.

Gravagnuolo, A., Girard, L. F., Ost, C., & Saleh, R. (2017). Evaluation criteria for a circular adaptive reuse of cultural heritage. *BDC. Bollettino Del Centro Calza Bini, 17*(2), 185–216.

Gunasingham, A. (2009, October 30). Singapore rejects emission cuts. *The Straits Times.*

Guzman, L. A., Peña, J., & Carrasco, J. A. (2020). Assessing the role of the built environment and sociodemographic characteristics on walking travel distances in Bogotá. *Journal of Transport Geography, 88.* https://doi.org/10.1016/j.jtrangeo.2020.102844

Hagan, S. (2014). *Ecological urbanism: The nature of the city.* Routledge.

Hamilton-Hart, N. (2006). Singapore's climate change policy: The limits of learning. *Contemporary Southeast Asia, 28*(3), 63–84.

Hamoud, M. (2007). The old town of Zanzibar and the struggle for reclaiming women's space in world heritage. *World Heritage and Identity: Three Worlds Meet*, 27–38.

Hanson, S., & Jones, A. (2015). Is there evidence that walking groups have health benefits? A systematic review and meta-analysis. *British Journal of Sports Medicine, 49*(11), 710–715.

Hausmann, A., Slotow, R., Fraser, I., & Di Minin, E. (2017). Ecotourism marketing alternative to charismatic megafauna can also support biodiversity conservation. *Animal Conservation, 20*(1), 91–100. https://doi.org/10.1111/acv.12292

Heinen, E., Van Wee, B., & Maat, K. (2010). Commuting by bicycle: An overview of the literature. *Transport Reviews, 30*(1), 59–96.

Henderson, J. C. (2013). Singapore's Little India: Ethnic districts as tourist attractions in B. Garrod & A. Fyall (Eds.), *Contemporary cases in heritage*, (pp. 34–50). Goodfellow.

Henderson, J. C. (2014). Bidding for World heritage: The case of Singapore's Botanic Gardens. *Tourism, Culture and Communication, 14*(2), 63–76.

Henderson, J. (2018). Making cities more walkable for tourists: A view from Singapore's streets. *International Journal of Tourism Cities, 4*(3), 285–297.

Herbig, U., Weichart, G., & Breuling, J. (2012). Gender roles and women's influence on traditional architecture and its development in Indonesia: a case study from Nias. In *Research & heritage: Research papers on architectural heritage* (pp. 263–287). The Saudi Commission for Tourism and Antiquities.

Heritage Alliance. (2019). *The heritage alliance, inspiring creativity, heritage & the creative industries a heritage alliance report*. https://ilucidare.eu/sites/default/files/2019-10/inspiringcreativity_thareport.pdf. Accessed on 4 December 2022.

Higginbottom, K. (2004). Wildlife tourism. *CRC for Sustainable. Tourism, 1*(1), 1–301.

Higgins-Desbiolles, F. (2011). Resisting the hegemony of the market: Reclaiming the social capacities of tourism. In S. McCabe, L. Minnaert, & A. Diekmann (Eds.), *Social tourism in Europe: Theory and practice*. Channel View Publications.

Hjorthol, R. (2013). Transport resources, mobility and unmet transport needs in old age. *Ageing and Society, 33*(7), 1190–1211. https://doi.org/10.1017/S0144686X12000517

Ho, A. (2015). *Sentosa*. https://eresources.nlb.gov.sg/infopedia/articles/SIP_247_2005-01-20.html. Accessed on 4 Feb 2022.

Ho, K. C. (2017). Era of globalization: Singapore's new urban economy and the rise of a World Asian City. In C. K. Heng (Ed.) *50 years of urban planning in Singapore*. World Scientific Publishing Corporation.

Ho, K. C., & Lim, N. E. V. (1992). *Backlanes as contested regions: Construction and control of physical space*. Singapore University Press.

Ho, A. (2009, October 30). Reasons for Singapore to be cool on global warming. *The Straits Times*.

Ho. P. (2021, October 4). Singapore and the Science of Cities. *The Straits Times*.

Honeycombers. (2021, March 8). *Going hiking in Singapore? Here are the best nature trails to trek across the island*. https://thehoneycombers.com/singapore/hiking-in-singapore-trekking/. Accessed on 17 October 2021.

Hollenhorst, S. J., Houge-MacKenzie, S., & Ostergren, D. M. (2014). The trouble with tourism. *Tourism Recreation Research, 39*(3), 305–319.

Hoon, H. T. (2019). Economic dynamism, openness. *and inclusion: How Singapore can make the transition from an era of catch-up growth to life in a mature economy*. World Scientific Publishing Corporation. https://doi.org/10.1142/10883

Housing and Development Board. (2013). *Landscape guide*. Development and Procurement Group.

Hudson, C. (2015). ION Orchard: Atmosphere and consumption in Singapore. *Virtual Communication, 14*(3), 289–308.

Hui, W. T., & Helfer, C. (2016). Migration policies: Lessons from the Singaporean and Swiss experiences. In Y. Guo & J. J. Woo (Eds.), *Singapore and Switzerland: Secrets to small state success*. World Scientific Publishing Corporation

Ibrahim, F. (2005). Attitudes to transport modes for shopping purposes in Singapore. *Transport Reviews, 25*(2), 221–243.

Ibrahim, F., & Lim, F. P. (2005). The development of neighborhood centers in Singapore: From traditional to cluster shopping. *Pacific Rim Property Research Journal, 11*(4), 373–392.

In.Corp (2022, October 22). *Why is Singapore a prime spot for top-tier foreign talent?* https://www.incorp.asia/blogs/why-singapore-prime-spot-top-foreign-talent/. Accessed on 5 March 2023.

International Labour Organization (ILO). (1996–2022). *Creating decent job opportunities through applying employment intensive approaches at cultural heritage conservation activities (project description)*. https://www.ilo.org/beirut/projects/WCMS_836858/lang--en/index.htm. Accessed on 23 July 2022.

Italy-Crotia.EU (2019). *MADE IN-LAND Management and Development of INLANDs* (project description). https://www.italycroatia.eu/web/madeinland. Accessed on 5 December 2022.

Jacobs, J. (1972). *The death and life of great American cities*. Penguin.

Jerke, D. (2008). *Urban design and the bottom line: Optimizing the return on perception*. ULI-the Urban Land Institute.

Jeuring, J., & Haartsen, T. (2017). The challenge of proximity: The (un)attractiveness of near-home tourism destinations. *Tourism Geographies, 19*(1), 118–141. https://doi.org/10.1080/14616688.2016.1175024

Jung, H. J., Lee, Y., Kim, H., & Yang, H. (2014). Impacts of country images on luxury fashion brand: Facilitating with the brand resonance model. *Journal of Fashion Marketing and Management, 18*(2), 187–205.

Kadisly S. (2017, April 6). Hidden waters blog. *Alexandra Canal Singapore*. https://hiddenwatersblog.wordpress.com/2017/04/06/alexcanal/#:~:text=Alexandra%20Canal%20makes%20its%20appearance,consort%20of%20King%20Edward%20VII. Accessed on 6 December 2022.

Kam, N. J. Y. (2018). *Understanding heritage in Singapore: The irony behind Chinatown, Little India and Kampong Glam.* https://dr.ntu.edu.sg/handle/10356/73594. Accessed on 4 December 2022.

Kaur, G. (2016, May 20). Street life thrives in car-free zones. *The Straits Times.* https://www.straitstimes.com/lifestyle/street-life-thrives-in-car-free-zones. Accessed on 25 October 2017.

Kearney, A. (2008). Intangible cultural heritage: Global awareness and local interest. *Intangible heritage* (pp. 223–240). Routledge.

Keller, K. L. (2001). Building customer-based brand equity. *Marketing Management, 10*(2), 14–19.

Khalid, C. (2021, May 21). *The best nature trails in Singapore.* https://www.timeout.com/singapore/things-to-do/the-best-nature-trails-in-singapore. Accessed on 17 October 2021.

Kiaka, R., Chikulo, S., Slootheer, S., & Hebinck, P. (2021). "The street is ours". A comparative analysis of street trading, Covid-19 and new street geographies in Harare, Zimbabwe and Kisumu, Kenya. *Food Security*, 1–19.

Kiew, R., & Chan, L. (2013). Habitat management: The strategy to preserve the biodiversity of the Singapore Botanic Gardens rain forest. *Plant, People and Planet Earth: The Role of Botanic Gardens in Sustainable Living,* National Botanic Institute, Kirstenbosch.

Kim, B., Lee, D. C., Chua, B. L., & Han, H. (2022). Country image of gastronomy and branding Hawker Culture: Local consumers' perception. *Tourism Management Perspectives*, *44*. https://doi.org/10.1016/j.tmp.2022.101036

Klimczuk, A. (2015). *Economic foundations for creative ageing policy.* Palgrave Macmillan.

Koh, P. P., & Wong, Y. D. (2013). Comparing pedestrians' needs and behaviours in different land use environments. *Journal of Transport Geography, 26*, 43–50.

Koh, W. T. H., & Lee, D. K. C. (1994). The vehicle quota system in Singapore: An assessment. Transportation research Part A: Policy and practice. *Research Collection Lee Kong Chian School Of Business. 28*(1), 31–47. https://ink.library.smu.edu.sg/cgi/viewcontent.cgi?referer=&httpsredir=1&article=4363&context=lkcsb_research#:~:text=The%20first%20COE%20auction%20was,auctions%20on%20a%20quarterly%20basis. Accessed on 26 February 2023.

Kok, Y. F. (2021, July 7). Grab to trial new green ride-hailing service. *The Straits Times.*

Koh, Y. F. (2021, August 31). Campaign launched in Sembawang to encourage residents to cycle safely as more take to it amid pandemic. *The Straits Times.*

Kok, Y. F. (2023, January 13). New cycling paths in 7 towns to be ready from 2024. *The Straits Times.*

Kosasih, J. (2019, September 20). *GP Extra Credit: Ten years after their construction, have the Integrated Resorts had a significant impact on Singapore society?* https://eunoiapress.wordpress.com/2019/09/20/gp-extra-cre dit-ten-years-after-their-construction-have-the-integrated-resorts-had-a-signif icant-impact-on-singapore-society/ Accessed on 27 January 2023.

Krambeck, H., & Shah, J. (2006). *The global walkability index: Talk the walk and walk the talk.* Paper presented at the Better Air Quality Conference (BAQ).

Kua, E. H. (2018). *Colours of ageing: 30 years of research on the mental health of the Singapore elderly*. Write Editions.

Kua, E. H. (2021). *Towards ageing well: Planning a future-ready Singapore.* Centre for Liveable Cities Singapore. https://www.clc.gov.sg/docs/def ault-source/urban-systems-studies/uss-towards-ageing-well.pdf. Accessed on 8 May 2022.

Küpers, W., & Wee, D. (2018). Tourist Cities as embodied places of learning: Walking in the "Feelds" of Shanghai and Lisbon. *International Journal of Tourism Cities, 4*(3), 376–390.

Kuttan, S. C. (2021, March 21). *Commentary: Taking Singapore's green ambitions to new heights with a circular economy.* https://www.channelnewsasia. com/commentary/singapore-circular-economy-sustainability-community-cli mate-315981. Accessed on 15 May 2022.

Kwek, M. (2021, March 20). *8 Kopitiam secrets shared by Hawker Centre aunties & uncles that you should know besides tissue Chope-ing*. https://the smartlocal.com/read/kopitiam-secrets/. Accessed on 12 September 2021.

Labadi, S., Giliberto, F., Rosetti, I., Shetabi, L., & Yildirim, E. (2021). *Heritage and the sustainable development goals: Policy guidance for heritage and development actors. Paris: ICOMOS—International Council on Monuments and Sites.* https://www.icomos.org/images/DOCUMENTS/Secret ariat/2021/SDG/ICOMOS_SDGs_Policy_Guidance_2021.pdf. Accessed on 5 December 2022.

Lai, K. (2021, May 27). *The ultimate list of 23 Hiking trails in Singapore—Ranked from chill to adventurous*. https://thetravelintern.com/hiking-trails-in singapore/. Accessed on 17 October 2021.

Lai, S., Loke, L. H., Hilton, M. J., Bouma, T. J., & Todd, P. A. (2015). The effects of urbanisation on coastal habitats and the potential for ecological engineering: A Singapore case study. *Ocean & Coastal Management, 103*, 78–85.

Laurent, B., Doganova, L., Gasull, C., & Muniesa, F. (2021). The test bed island: Tech business experimentalism and exception in Singapore. *Science as Culture, 30*(3), 367–390.

Lee, H. H. (2019). *Analysis of walkability estimates and hotel real estate values in New York City*. https://ecommons.cornell.edu/bitstream/handle/1813/71332/Hyun_Ho_Lee_Latin_Honors_Thesis_Paper.pdf?sequence=1&isAllowed=y. Accessed on 2 February 2022.

Lee, K. Y. (1998). *The Singapore Story*. Singapore Press Holdings and Times Edition.

Lee, M. T. (2022, June 18). Car aspirations wane: Few are willing to commute on bicycle. *The Straits Times*.

Lee, M. T. G. (2000). *The making of Jurong Island—The right chemistry*. Jurong Town Corporation.

Lee, S. K., & Chua, S. E. (1992). *More than a garden city*. Singapore: Parks & Recreation Dept., Ministry of National Development. Call No.: RSING 333.783095957 LEE, 13.

Lee, S. T., & Kim, H. S. (2021). Food fight: Gastrodiplomacy and nation branding in Singapore's UNESCO bid to recognize Hawker culture. *Place Branding and Public Diplomacy, 17*(2), 205–217.

Leong, W., Goh, K., Hess, S., & Murphy, P. (2016). Improving bus service reliability: The Singapore experience. *Research in Transportation Economics, 59*, 40–49.

Lerario, A. (2022). The role of built heritage for sustainable development goals: From statement to action. *Heritage, 5*(3), 2444–2463.

Levine, R. (1997). *A geography of time: The temporal misadventures of a social psychologist, or how every culture keeps time just a little bit differently*. Basic Books.

Liew, C. (2022). Will car-sharing replace car ownership in Singapore?. https://vibrantdot.co/will-car-sharing-replace-car-ownership-in-singapore/. Accessed on 21 May 2022.

Lin, S. L. (2018). "We work like ants… we avoid being troublemaker": An exploratory inquiry on resilience of Chinese street vendors in the urban village. *International Journal of Sociology and Social Policy., 38*(11/12), 1024–1040. https://doi.org/10.1108/IJSSP-01-2018-0008

Lingle, C. (1996). *Singapore's authoritarian capitalism: Asian values, free market illusions, and political dependency*. Edicions Sirocco, S.L.

Litman, T. (2017). Economic value of walking. In *Walking*. Emerald Publishing Limited, 81–98. Accessed on 19 September 2021.

Lo, R. H. (2009). Walkability: What is it? *Journal of Urbanism, 2*(2), 145–166.

Lopez, M. C. R., & Wong, Y. D. (2017). Attitudes towards active mobility in Singapore: A qualitative study. *Case Studies on Transport Policy, 5*(4), 662–670.

Louie, J. (2018, March 19). *Understanding the culture of Malacca's Kristang community*. https://www.destinasian.com/blog/news-briefs/malacca-kristang-community. Accessed on 1 January 2023.

Low, J., & Lim, L. J. (2012). Conservation of our natural heritage: The Singapore experience. In *Proceedings of the 12th International Coral Reef Symposium, Cairns, Australia* (pp. 10–13).

LTA. (2021). *Greener commutes, sustainable lifestyles annual report*. Singapore: LTA.

Lumsdon, L. (2000). Transport and tourism: Cycle tourism—A model for sustainable development? *Journal of Sustainable Tourism, 8*(5), 361–377. https://doi.org/10.1080/09669580008667373

Lye, L. H. (2008, July) A fine city in a garden—environmental law and governance in Singapore. *Singapore Journal of Legal Studies*, 68–117.

Ma, L., & Guek, L. (2012). *A delightful eco-nature neighbourhood park*. CityGreen #6, Centre for Urban Greenery and Ecology Publication, CUGE.

Mackett, R. L., & Brown, B. (2011). *Transport, physical activity and health: Present knowledge and the way ahead*. https://discovery.ucl.ac.uk/id/eprint/1333502/. Accessed on 1 October 2021.

Mahbubani, K. (2015). *Can Singapore survive?* Straits Times Press.

Mahbubani, K. (2017). Singapore: the smartest city on our planet. In Heng, C. K. (Ed.) *50 years of urban planning in Singapore*. World Scientific Publishing Corporation.

Mansfield, T. J., & MacDonald Gibson, J. (2015). Health impacts of increased physical activity from changes in transportation infrastructure: Quantitative estimates for three communities. *BioMed Research International*. https://doi.org/10.1155/2015/812325

Marshall, S. (2004). *Streets and patterns*. Routledge.

McAnirlin, O., & Maddox, C. B. (2022). 'We have to be a little more realistic': Women's outdoor recreation experiences in a community hiking group. *Annals of Leisure Research, 25*(3), 335–351.

Means, G. (1996). Soft authoritarianism in Malaysia and Singapore. *Journal of Democracy, 7*(4), 103–117.

Mediacorp Pte Ltd. (2022). *About CNA Green Plan*. https://www.channelnewsasia.com/cna-green-plan. Accessed on 21 May 2022.

Mejabi, E. I. (2018). *The impact of events on sustainable tourism development in Durban: An evidence-based enquiry* (Doctoral dissertation).

Ministerial Committee on Ageing. (n.d.-a). *What is the action plan about? I feel young SG—MOH*. https://www.moh.gov.sg/ifeelyoungsg/about/what-is-the-action-plan-about. Accessed on 8 May 2022).

Ministerial Committee on Ageing (n.d.-b). *Go about easily*. https://www.moh.gov.sg/ifeelyoungsg/how-can-i-age-in-place/go-about-easily#:~:text=A%20SENIOR%2DFRIENDLY%20TRANSPORT%20NETWORK&text=Senior%20citizens%20with%20concession%20cards,the%20lower%20pre%2Dpeak%20fares.. Accessed on 26 Feb 2023).

Ministry of Education, Singapore. (2014). *Applied study in polytechnics and ITE review (ASPIRE) report*. Singapore: Ministry of Education.

Ministry of the Environment and Water Resources. (2009). *A lively and liveable Singapore: Strategies for sustainable growth*. Ministry of the Environment and Water Resources and Ministry of National Development.

Ministry of Foreign Affairs Singapore. (2018). *Towards a sustainable and resilient Singapore*. https://sustainabledevelopment.un.org/content/documents/19439Singapores_Voluntary_National_Review_Report_v2.pdf. Accessed on 8 January 2023.

Ministry of National Development, Planning Department. (1975). *Revised master plan: Report of survey* (pp. 68–69). The Department. Call No.: RCLOS 711.4095957 SIN.SMNDAR.

Ministry of National Development. (2007). *From garden city to city in a garden*. http://www.mnd.gov.sg/MNDAPPImages/About%20Us/From%20Garden%20City%20to%20City%20in%20a%20Garden.pdf. Accessed on 12 February 2015.

Ministry of Transport. (n.d.). *Sustainable Land transport policies*. https://www.mot.gov.sg/what-we-do/green-transport/sustainable-land-transport. Accessed on 25 March 2023.

Monteiro, R., Ferreira, J. C., & Antunes, P. (2020). Green infrastructure planning principles: An integrated literature review. *Land*, *9*(12). https://doi.org/10.3390/land9120525

Mooney, C., & Freedman, A. (2021, January 25). Earth is now losing 1.2 trillion tons of ice each year. And it's going to get worse. *Washington Post*. https://www.washingtonpost.com/climate-environment/2021/01/25/ice-melt-quickens-greenland-glaciers/. Accessed on 14 January 2023.

Murray, L. (2015). Age-friendly mobilities: A transdisciplinary and intergenerational perspective. *Journal of Transport & Health, 2*(2), 302–307. https://doi.org/10.1016/j.jth.2015.02.004

Mutalib, H. (2000). Illiberal democracy and the future of opposition in Singapore. *Third World Quarterly, 21*(2), 313–342.

Naselli, F. (2016). Tourism as Resource in managing the valorisation of the urban and territorial heritage within the mediterranean basin. MOTRIS a Proposal of integrated relational tourism. *Procedia-Social and Behavioral Sciences, 225*, 433–438.

National Computer Board. (1992). *A vision of an intelligent island: The IT2000 Report*. Singapore: National Computer Board.

National Heritage Board. (2020, November). *Temple cluster at tampines link and the former Hun Yeang Village*. https://www.roots.gov.sg/places/places-landing/Places/landmarks/tampines-heritage-trail-green-spaces-trail/Temple-Cluster-at-Tampines-Link-and-the-former-Hun-Yeang-Village

National Heritage Board. (2022, July 12). *National monuments*. https://www.roots.gov.sg/nhb/monuments. Accessed on 20 November 2022.

National Heritage Board. (2023a). *Former ford factory*. https://www.roots.gov.sg/places/places-landing/Places/national-monuments/former-ford-factory. Accessed on 1 January 2023.

National Heritage Board. (2023b). *Istana Kampong Gelam*. https://www.roots.gov.sg/places/places-landing/Places/national-monuments/istana-kampong-glam. Accessed on 1 January 2023.

National Library Board, Singapore. (2019, July 31). *To wreck or to recreate: Giving new life to Singapore's built heritage*. https://biblioasia.nlb.gov.sg/vol-15/issue-2/jul-sep-2019/preservation-buildings. Accessed on 29 December 2022.

National Parks Board. (2020, April). *Recreational connectivity*. https://www.nparks.gov.sg/gardens-parks-and-nature/park-connector-network. Accessed on 6 June 2021.

National Parks Board. (2023). *Empress place*. https://www.nparks.gov.sg/gardens-parks-and-nature/parks-and-nature-reserves/empress-place. Accessed on 1 January 2023.

National University of Singapore (2020 January). *An assessment framework for monitoring cities' sustainability*. https://cde.nus.edu.sg/csac/wp-content/uploads/sites/29/2020/01/Full-template-for-Benchmark-Booklet.pdf. Accessed on 24 January 2023.

Newman, P. (2014). Biophilic urbanism: A case study on Singapore. *Australian Planner, 51*(1), 47–65.

Newman, P., & Kenworthy, J. (2015). *The end of automobile dependence* (pp. 105–140). Island Press/Center for Resource Economics.

Newsdesk, Inside Asian Geming. (2020, November 2). *Singapore regulator says casino entry levy increase has kept locals away*. https://www.asgam.com/index.php/2020/11/02/singapore-regulator-says-casino-entry-levy-increase-has-kept-singapore-locals-away/. Accessed on 28 January 2023.

Newton, P. W. (2007). Horizon 3 planning: Meshing liveability with sustainability. *Environment and Planning B: Planning and Design, 34*(4), 571–575.

Newton, P., Newman, P., Glackin, S., & Trubka, R. (2012). Greening the greyfields. *International Journal of Social, Behavioral, Educational, Economic, Business and Industrial Engineering, 6*(11), 2870–2889.

Ng, C. S. L., & Chou, L. M. (2018). Coral reef restoration in Singapore—Past, present and future. In *Sustainability Matters: Environmental Management in the Anthropocene*. (pp. 3–23).

Ng, K.G. (2022, May 31). 113 people under investigation for loan shark activities after week-long blitz. *The Straits Times*.

Ng, W. K. (2022, November 28). New CBD cycling paths link workplaces, MRT stations. *The Straits Times*.

Ng, W. S., & Mendelsohn, R. (2005). The impact of sea level rise on Singapore. *Environment and Development Economics, 10*(2), 201–225.

Niu, S., Hu, A., Shen, Z., Huang, Y., & Mou, Y. (2021). Measuring the built environment of green transit-oriented development: A factor cluster analysis of rail station areas in Singapore. *Frontiers of Architecture Research, 10*, 652–668.

Nordbakke, S., & Schwanen, T. (2014). Well-being and mobility: A theoretical framework and literature review focusing on older people. *Mobilities, 9*(1), 104–129. https://doi.org/10.1080/17450101.2013.784542

NParks. (2022a). *Esplanade park*. https://www.nparks.gov.sg/gardens-parks-and-nature/parks-and-nature-reserves/esplanade-park. Accessed on 22 December 2022.

NParks. (2022b). *OneMillionsTrees movement*. https://www.nparks.gov.sg/treessg/one-million-trees-movement. Accessed on 23 February 2023.

NParks. (2022c). *Sustainability report*. https://www.nparks.gov.sg/portals/annualreport/sustainability-report/index.htm. Accessed on 25 January 2023.

NParks Rail Corridor. (2023, April 3). *Our Rail corridor*. https://railcorridor.nparks.gov.sg/. Accessed on 8 April 2023.

Nur Dianah Suhaimi (2009, October 31). Emissions cut if others do their part as well. *The Straits Times*.

OECD. (n.d.). *OECD ICOM Guide for Cities and Museums on Culture and Local Development*. In Proceedings of the OECD Conference on Culture and Local Development, Venice, Italy. (6–7 December 2018). https://www.oecd.org/cfe/leed/venice-2018-conference-culture/. Accessed on 19 November 2022.

Olafsdottir, G., Cloke, P., Schulz, A., Van Dyck, Z., Eysteinsson, T., Thorleifsdottir, B., & Vögele, C. (2020). Health benefits of walking in nature: A randomized controlled study under conditions of real-life stress. *Environment and Behavior, 52*(3), 248–274.

Oliver, T. (2020). *The self-delusion: The surprising science of our connection to each other and the natural world*. Weidenfeld & Nicolson.

Olszewski, P., & Wibowo, S. S. (2005). Using equivalent walking distance to assess pedestrian accessibility to transit stations in Singapore. *Transportation Research Record: Journal of the Transportation Research Board, 1927*(1), 38–45.

Ong, B. L. (2003). Green plot ratio: An ecological measure for architecture and urban planning. *Landscape and Urban Planning, 63*(4), 197–211.

Onorati, M. G., & d'Ovidio, F. D. (2022). Sustainable eating in the "new normal" Italy: Ecological food habitus between biospheric values and de-globalizing gastronationalism. *Food, Culture & Society*, 1–20.

Ooi, G. L. (2004). *Future of Space*. Eastern University Press.

Ooi, G. L., & Kog, Y. C. (1999). Future Urbanisation. In Low, L (Eds.), *Singapore: Towards a developed status*. Centre for Advanced Studies and Oxford University Press.

Open Government Products. (2022, April 19). *Circular economy*. Circular Economy (towardszerowaste.gov.sg). Accessed on 15 May 2022.

Ortmann, S. (2011). *The 2011 elections in Singapore: The emergence of a competitive authoritarian regime*. Working Paper No. 112, City University of Hong Kong, November 2011.

Richards, G., & Wilson, J. (2006). Developing creativity in tourist experiences: A solution to the serial reproduction of culture? *Tourism Management, 27*(6), 1209–1223. https://doi.org/10.1016/j.tourman.2005.06.002

Pape, G. (2014). *To attract talent think what kind of place the emerging generation of professionals want to live*. https://www.canr.msu.edu/news/to_attract_talent_think_what_kind_of_place_the_emerging_generation_of_profe. Accessed 10 June 2022.

Pappu, R., Quester, P. G., & Cooksey, R. W. (2007). Country image and consumer-based brand equity: Relationships and implications for international marketing. *Journal of International Business Studies, 38*(5), 726–745.

Peerapun, W., Sereerat, S., Sanit, P., & Vichienpradit, P. (2020). Master planning for conservation and development of Krung Rattanakosin 2032. *Nakhara: Journal of Environmental Design and Planning, 19*, 39–58.

Perrotti, D. (2020). Toward an agentic understanding of the urban metabolism: A landscape theory perspective. *Urban Geography*, 1–11. https://doi.org/10.1080/02723638.2020.1848760

PUB. (2022). *Active, beautiful, clean waters programme*. https://www.pub.gov.sg/abcwaters/about. Accessed on 18 May 2022.

Public Works Department. (1971). *Annual report 1970* (p. 27). Singapore: Govt. Print. Off. Call no.: RCLOS 354.59570086 SIN, 27.

Public Works Department. (1972). *Annual report 1971* (p. 63). Singapore: Govt. Print. Off. Call no.: RCLOS 354.59570086 SIN, 63

Public Works Department. (1975). *Trees are inspected regularly* (p. 52). Tee, S. P. (2014, June 23). Trees are inspected regularly. *Today*. http://tablet.todayonline.com/voices/trees-are-inspected-regularly

Quah, J. (1979). Police corruption in Singapore: An analysis of its forms, extent and causes. *Singapore Police Journal, 10*, 7–43.

Quah, J. (2008). Good governance, accountability and administrative reform in Singapore. *American Journal of Chinese Studies, 15*(1), 17–34.

Quinn, B., & Wilks, L. (2013). Festival connections: People, place and social capital. In G. Richards, M. P. de Brito, & L. Wilks (Eds.), *Exploring the social impacts of events*. Routledge.

Reuters. (2016, July 23). Singapore street stalls make dining history with Michelin star. *Reuters*. https://www.reuters.com/article/us-singapore-food-hawkers-michelin-star-idUKKCN1021XA. Accessed on 18 December 2022.

Rockower, P. S. (2012). Recipes for gastrodiplomacy. *Place Branding and Public Diplomacy, 8*(3), 235–246.

Roubini, N. (2022). *Megathreats: The ten threats that imperil our future, and how to survive them*. John Murray (Publishers).

Rundle, G., Bader, M., Richards, C. A., Neckerman, K. M., & Teitler, J. O. (2011). Using Google street view to audit neighbourhood environments. *American Journal of Preventive Medicine, 40*(1), 94–100.

Samra, H. A, & Specker, B. (2007). Walking age does not explain term versus preterm difference in bone geometry. *The Journal of pediatrics*, *151*(1), 61–66. https://doi.org/10.1016%2Fj.jpeds.2007.02.033

Samarasekara, G. N., Fukahori, K., & Kubota, Y. (2011). Environmental correlates that provide walkability cues for tourists: An analysis based on walking decision narrations. *Environment and Behaviour, 43*(4), 501–524.

Sarmento, J. (2016). Tourists' walking rhythms: 'Doing' the Tunis Medina. *Tunisia, Social & Cultural Geography, 18*(3), 295–314.

Savage, V. (1997). Singapore's garden city: Translating Environmental possibilism. In G. L. Ooi, G. L., & K. Kowk (Eds.), *City and the state: Singapore's Built environment revisited*. The Institute of Policy Studies and Oxford University Press.

Saxena, N. C. (2011). *Virtuous cycles: The Singapore public service and national development*. United Nations Development Programme.

Schein, E. H. (1996). *Strategic pragmatism*. The MIT Press.

Schlossberg, M., Johnson-Shelton, D., Evers, C., & Moreno-Black, G. (2015). Refining the grain: Using resident-based walkability audits to better understand walkable urban form. *Journal of Urbanism, 8*(3), 260–278.

Seoul Institute. (2016). *Creating walkable and bikeable cities: Lessons from Seoul and Singapore*. https://www.clc.gov.sg/docs/default-source/books/walkable-and-bikeable-cities.pdf. Accessed on 30 October 2021.

SG101. (2022). *Preserving a Unique national identity*. https://www.sg101.gov.sg/infrastructure/urban-planning/unique#:~:text=In%201971%2C%20the%20Government%20passed,eight%20buildings%20under%20its%20protection. Accessed on 29 December 2022.

SG Press Centre. (2012, July 1). *Speech by Mr. Lee Hsien Loong*. World Cities Summit, Singapore.

Shatkin, G. (2014). Reinterpreting the meaning of the 'Singapore model': State capitalism and urban planning. *International Journal of Urban and Regional Research, 38*(1), 116–137.

Silver Travel Advisor. (2022). *Silver Travel advisor industry snapshot April 2022.* https://en.calameo.com/silvertraveladvisor/read/0044274231677a12c8b3f. Accessed on 26 December 2022.

Singapore Cooperation Programme. (March 9, 2022). *Integrated urban transport planning and management.* https://scp.gov.sg/startpublic/#!/courses/Search/transport/8685/Integrated%20Urban%20Transport%20Planning%20and%20Management/5/CLS_3. Accessed on 26 March 2023.

Singapore Department of Statistics. (2019). *Population trends 2019.* https://www.singstat.gov.sg/-/media/files/publications/population/population2019.pdf. Accessed on 8 May 2022.

Singapore Department of Statistics. (2021). *Population trends 2021.* https://www.singstat.gov.sg/-/media/files/publications/population/population2021.ashx. Accessed on 8 May 2022.

Singapore Economic Development Board. (2021, August 25). *A great place to live in.* https://www.edb.gov.sg/en/why-singapore/a-great-place-to-live.html. Accessed on 4 September 2021.

Singapore Green Plan 2030. (2021). *Singapore Green Plan 2030a charts ambitious targets for next 10 years to catalyse national sustainability movement.* www.greenplan.gov.sg/resource-room/2021-02-10-press-release-on-green-plan. Accessed on 18 December 2022.

Singapore Green Plan 2030. (2022). *Our global commitment to climate change and sustainable development.* https://www.greenplan.gov.sg/globalcommitment. Accessed on 8 January 2023.

Singapore Green Plan 2030. (2022). *Our sustainable development movement for a better and greener future.* https://www.greenplan.gov.sg/overview. Accessed on 27 December 2022.

Singapore Management University. (2019, March 25). University builds bridges to innovation. *The Straits Times.* chrome-extension://efaidnbmnnnibpcajpcglclefindmkaj/https://news.smu.edu.sg/sites/news.smu.edu.sg/files/wwwsmu/news_room/smu_in_the_news/2019/Mar2019/Mar25/20190325-STHM-Home-B06-33x41.pdf. Accessed on 7 April 2023.

Singapore Tourism Board. (2016). *Marketing strategy: Of stories, fans, and channels.* https://www.stb.gov.sg/content/dam/stb/documents/assistancelicensing/marketing-innovation/Marketing_Strategy-Of_Stories_Fans_and_Channels.pdf. Accessed on 26 December 2022.

Singapore Tourism Board. (2022). *Eurasian Heritage Centre.* https://www.visitsingapore.com/en_my/see-do-singapore/culture-heritage/heritage-discovery/eurasian-heritage-centre/. Accessed on 1 January 2023.

Singapore Tourism Board. (2023). *Faces of Singapore.* https://www.visitsingapore.com/travel-guide-tips/about-singapore/people-of-singapore/. Accessedon 12 January 2023.

Singapore Zoological Gardens. (2022). *Our role in conservation.* https://www.mandai.com/en/protecting-wildlife/conservation.html. Accessed on 2 February 2022.

Smart Nation and Digital Government Office. (2018). *Smart nation: The Way forward.* Singapore: Smart Nation and Digital Government Office.

Smith, M. K., & Diekmann, A. (2017). Tourism and wellbeing. *Annals of Tourism Research, 66*, 1–13. https://doi.org/10.1016/j.annals.2017.05.006

Soh, S. H. (2000). *Singapore. In Asian approach to resource conservation and environmental protection.* Asian Productivity Organization.

Spencer, A. H. (1988). Modernisation and Incorporation: The Development of Singapore's Bus Services 1945–1974. *Environment and Planning a: Economy and Space, 20*(8), 1027–1046. https://doi.org/10.1068/a201027

SPH Media Limited. (2022, April 24). *Recycling efforts shouldn't go to waste.* https://www.straitstimes.com/opinion/st-editorial/recycling-efforts-shouldnt-go-to-waste. Accessed on 21 May 2022.

Stanley, P. (2020). Unlikely hikers? Activism, Instagram, and the queer mobilities of fat hikers, women hiking alone, and hikers of colour. *Mobilities, 15*(2), 241–256.

StayNTouch (2022). *2022 Hotelier technology sentiment report.* https://www.stayntouch.com/2022-hotelier-technology-sentiment-report-2/. Accessed on 8 April 2023.

Subramaniam, M., Chong, S. A., Vaingankar, J. A., Abdin, E., Chua, B. Y., Chua, H. C., Eng, G.K., Heng, D., Hia, S.B., Huang, W. and Jeyagurunathana, A., Kua, J., Lee, S. P., Mahendran, R., Magadi, H., Malladi, S., McCrone, P., Pang, S., Picco, L., Sagayadevan, V., Sambasivam, R., Seng, K. H., Seow, E., Snafie, S., Shahwan, S., Tang, L. L., Yap, M., Zhang, Y., Ng, L. L., & Prince, M. (2015). Prevalence of dementia in people aged 60 years and above: Results from the WiSE study. *Journal of Alzheimer's Disease, 45*(4), 1127–1138.

Sun, J. (2015). *Looking to the future designing and managing residential facilities for people with Dementia.* Agency for Integrated Care.

Szubski, C. (2017, December). *Singapore—A High-density city with sporting characters.* https://sportifycities.com/singapore-cycling/. Accessed on 19 September 2021.

Szubski, C. (2022). *Singapore: Cycling.* https://sportifycities.com/singapore-cycling/. Accessed on 26 February 2023.

Tan, E., & Abu Bakar, B. (2018). Ethnic reunions in tourism: The route to roots. In *Asian cultures and contemporary tourism* (pp. 21–39).

Tan, E. C. (2020, September 10). *Universities need to tear down subject silos.* https://www.straitstimes.com/opinion/universities-need-to-tear-down-subject-silos. Accessed on 7 April 2023.

Tan, A. G. L., & Daud, N. I. (2020). Street trading activities in Kuala Lumpur's Chinatown: Present situation and future directions. *Malaysia Architectural Journal, 2*(1), 38–48.

Tan, E., & Law, R. (2016). mLearning as a softer visitor management approach for sustainable tourism. *Journal of Sustainable Tourism, 24*(1), 132–152. https://doi.org/10.1080/09669582.2015.1049610

Tan, K. (2022). *A history of the padang*. https://biblioasia.nlb.gov.sg/vol-18/issue-1/apr-to-jun-2022/history-padang. Accessed on 22 December 2022.

Tao, T., Wu, X., Cao, J., Fan, Y., Das, K., & Ramaswami, A. (2020). Exploring the nonlinear relationship between the built environment and active travel in the twin cities. *Journal of Planning Education and Research*. https://doi.org/10.1177%2F0739456X20915765

Tay, J. Y., Wong, S. K., Chou, L. M., & Todd, P. A. (2018). Land reclamation and the consequent loss of marine habitats around the Ayer Islands, Singapore. *Nature in Singapore, 11*, 1–5.

Taylor, J. (2015). Embodiment unbound: Moving beyond divisions in the understanding and practice of heritage conservation. *Studies in Conservation, 60*(1), 65–77.

Te Brommelstroet, M., Nikolaeva, A., Glaser, M., Skou Nicolaisen, M., & Chan, C. (2017). Travelling together alone and alone together: Mobility and potential exposure to diversity. *Applied Mobilities, 2*(1), 1–15. https://doi.org/10.1080/23800127.2017.1283122

Teo, A. (2020, October). *Starting a circular economy in Singapore from zero waste*. https://www.clc.gov.sg/research-publications/publications/digital-library/view/circular-economy-in-singapore-from-zero-waste. Accessed on 15 May 2022.

Teo, P., Mehta, K., Thang, L. L., & Chan, A. (2006). *Ageing in Singapore: Service needs and the state*. Routledge.

Tham, Y. C. (2022, January 25). Singapore has to plan for more flexibility in the city: Desmond Lee. *The Straits Times*.

The Economist. (2022, February 21). Thailand's gastro-diplomacy. https://www.economist.com/asia/2002/02/21/thailands-gastro-diplomacy. Accessed on 18 December 2022.

The Explorer. (2020, March 12). How the circular economy is changing business. https://www.theexplorer.no/stories/renewable-resources/an-introduction-to-the-circular-economy/. Accessed on 17 May 2022.

The European Parliament and the Council. (2010, May 19). *Directive on the Energy Performance of Buildings*. https://energy.ec.europa.eu/topics/energy-efficiency/energy-efficient-buildings/energy-performance-buildings-directive_en . Accessed on 4 December 2022.

The Straits Times. (1969, January 23). A 'Paradise Island'. p. 4. https://eresources.nlb.gov.sg/newspapers/Digitised/Article/straitstimes19690123.2.20.. Accessed on 4 February 2022.

The Straits Times. (1972, March 24). Sentosa pleasure island will have Pirates Cove. p. 32 . https://eresources.nlb.gov.sg/newspapers/Digitised/Article/straitstimes19720324-1.2.110. Accessed on 4 February 2022.

The Straits Times. (2022, May 31). 113 people under investigation for loan shark activities after week-long blitz. https://www.straitstimes.com/singapore/courts-crime/113-persons-under-investigation-for-loan-shark-activities-after-week-long-blitz. Accessed on 28 January 2023.

Thomas, L., & Cousins, W. (1996). *A new compact city form: Concepts in practice* (pp. 328–338). A sustainable urban form, Spon Press.

Thomson, G., & Newman, P. (2016). Geoengineering in the Anthropocene through regenerative urbanism. *Geosciences*, *6*(4), https://doi.org/10.3390/geosciences6040046

Toh, A. Q. W., Chee, W. J., Lai, J. Q., Aziz-Boey, N. M., Tan, H. F., Chester, M., & Chiam, V. H. T. (2017). *Safeguarding intangible cultural heritage in Chinatown: Threats & strategies, policy analysis exercise*. Lee Kuan Yew School of Public Policy. https://www.singaporeheritage.org/wp-content/uploads/2018/05/LKYSPP-SHS-Cultural-Heritage-in-Chinatown-Final-Report-13-Nov-2017.pdf. Accessed on 6 December 2022.

Toh, T. W. (2021a, December 19). App helps cyclists plan routes on paths, park connectors. *The Straits Times*.

Toh, T. W. (2021b, December 20). 50 roads to be assessed if parts can be spared for cycling paths. *The Straits Times*.

Tortajada C., & Biswas, A. K. (2017, December 31). What made Singapore tick: soft option and hard rules. *The Straits Times*.

Trading Economics. (2022). *Singapore—Land Area (sq. km)*. https://tradingeconomics.com/singapore/land-area-sq-km-wb-data.html. Accessed on 14 January 2023.

Trading Economics. (2023). *Singapore GDP from transport*. https://tradingeconomics.com/singapore/gdp-from-transport. Accessed on 26 February 2023.

Turnbull, C. M. (1972). *The Straits settlements 1826–67: Indian Presidency to crown colony*. Athlone Press.

UNESCO Institute of Statistics. (2009). *Culture heritage*. https://uis.unesco.org/en/glossary-term/cultural-heritage. Accessed on 29 December 2022.

UNESCO Intangible Cultural Heritage. (1992–2022). *Hawker culture in Singapore, community dining and culinary practices in a multicultural urban context*. https://ich.unesco.org/en/RL/hawker-culture-in-singapore-community-dining-and-culinary-practices-in-a-multicultural-urban-context-01568. Accessed on 1 January 2023.

UNESCO World Heritage Centre. (1992–2022). *Singapore botanic gardens.* https://whc.unesco.org/en/list/1483/. Accessed on 4 December 2022.

United Cities and Local Governments. (2010, November 17). *Culture: Fourth pillar of sustainable development*, (Statement). https://www.agenda21culture.net/sites/default/files/files/documents/en/zz_culture4pillarsd_eng.pdf. Accessed on 19 November 2022).

United Institute of Statistics. (2023). *Sustainable development goal 11.4.* https://uis.unesco.org/en/topic/sustainable-development-goal-11-4. Accessed on 8 January 2023.

United Nations Climate Change. (n.d.). *The Paris Agreement. What is the Paris Agreement?* https://unfccc.int/process-and-meetings/the-paris-agreement. Accessed on 27 March 2023.

United Nations Conference on Environment and Development. (1992). *Singapore's national report for the 1992 UN Conference on Environment and Development Preparatory Committee* (United Nations: Inter-Ministry Committee for the UNCED Preparatory Committee).

United Nations-Habitat. (2013). *State of the World's Cities 2012–2013 report.* https://unhabitat.org/sites/default/files/download-manager-files/State%20of%20the%20World%20Cities%2020122013.pdf. Accessed on 19 November 2022.

United Nations—Habitat III. (2016). *New Urban Agenda.* https://unhabitat.org/sites/default/files/2019/05/nua-english.pdf. Accessed on 4 May 2022.

Urban Redevelopment Authority. (2022, May 18). *Master Plan.* https://www.ura.gov.sg/Corporate/Planning/Master-Plan. Accessed on 18 May 2022.

Urban Redevelopment Authority. (2022). *Protecting & Rejuvenating Built heritage.* https://www.ura.gov.sg/Corporate/Planning/Master-Plan/Themes/Rejuvenating-Familiar-Places/Protecting-and-Rejuvenating-Built-Heritage. Accessed on 22 November 2022.

Urban Redevelopment Authority. (2023). *Heritage impact assessment.* https://www.ura.gov.sg/Corporate/Planning/Our-Planning-Process/Bringing-plans-to-Reality/Heritage-Impact-Assessment. Accessed on 24 January 2023.

Urban Redevelopment Authority of Singapore Centre for Liveable Cities. (2016, May 5–7). *Heritage and Sustainable urbanism: Case Studies from France, Singapore And the region.* https://www.clc.gov.sg/docs/default-source/reports/heritage-symposium-report.pdf. Accessed on 5 December 2022.

Verster, M. (2014). *The wall and the veil: reclaiming women's space in a world heritage site* (Doctoral dissertation, University of Pretoria).

VinHMS. (2022). *Millennials travel trends redefine the hospitality in 2022.* https://www.cihms.com/en/millennials-travel-trends-redefine-the-hospitality-in-2022/#:~:text=Experiential%20travel&text=An%20Expedia%20survey%20has%20found,at%20premium%20hotels%20and%20resorts. Accessed on 26 December 2022.

Wahl, H. W., & Lang, F. R. (2003). Aging in context across the adult life course: Integrating physical and social environmental research perspectives. *Annual Review of Gerontology and Geriatrics, 23*, 1–33.
Walker, R. A. (2007). *The country in the city: The greening of the San Francisco Bay area. Weyerhaeuser Environmental Books*. University of Washington Press.
Wang, L., & Wen, C. (2017). The relationship between the neighborhood-built environment and active transportation among adults: A systematic literature review. *Urban Science 1*(3). https://doi.org/10.3390/urbansci1030029
Widjajanti, R., & Ardiati, H. H. (2020). Street vendors' activity space in the residential, public space (Case Study: Tirto Agung Park, Semarang). In *IOP Conference Series: Earth and Environmental Science, 409*(1), IOP Publishing. https://doi.org/10.1088/1755-1315/409/1/012029
Wildermuth, B. (2016). Land transport policy: Urban infrastructure in Singapore and Switzerland. In Y. Guo & J. J. Woo (Eds.), *Singapore and Switzerland: Secrets to small state success*. World Scientific Publishing Corporation.
Willems, T., & Graham, C. (2019). The imagination of Singapore's Smart Nation as digital infrastructure: Rendering (digital) work invisible. *East Asian Science, Technology and Society: An International Journal, 13*(4), 511–536.
Williams, B. (2022, December 9). *80+ Family Travel & tourism statistics (2022)*. https://www.dreambigtravelfarblog.com/blog/family-travel-tourism-statistics. Accessed on 26 December 2022.
Wilson, E. (1984). *Biophilia*. Harvard University Press.
Wong, T. C. (2008). Integrated resort in the central business district of Singapore: The land use planning and sustainability issues. *Spatial planning for a sustainable Singapore*, 59–78.
Worldometer. (n.d.). *Life expectancy of the world population*. https://www.worldometers.info/demographics/life-expectancy/. Accessed on 8 May 2022.
World Health Organization. (2021, June 21). *Road traffic injuries*. https://www.who.int/news-room/fact-sheets/detail/road-traffic-injuries. Accessed on 26 August 2021.
World Resources Institute (WRI) and Wbscd. (n.d.). *Greenhouse gas protocol*. https://ghgprotocol.org/. Accessed on 19 May 2022.
World Travel & Tourism Council. (2022). *Economic impact report*. https://wttc.org/research/economic-impact. Accessed on 8 April 2023.
Wu, T. (2021). *7 Best Hawker Centres you should visit in Singapore*. https://bestinsingapore.com/hawker-centres/. Accessed on 10 September 2021.
Yeoh, B. S. A. (2003). *Contesting space in Colonial Singapore: Power relations and the urban built environment*. Singapore University Press.
Yeo, D. (2013, January 7). *Making Singapore a city for all ages*. Challenge. https://www.psd.gov.sg/challenge/ideas/deep-dive/making-singapore-a-city-for-all-ages. Accessed on 11 May 2022.
Yeo, G. (2022). *Musings (Series One)*. World Scientific.

Yok, T. P. (2012). *Vertical garden city, Singapore*. https://www.greenroofs.com/2013/11/27/vertical-garden-city-singapore-by-tan-puay-yok/. Accessed on 16 May 2022.

Yong, C. (2021, November 6). Stretch of civic district to be fully for pedestrians. *The Straits Times*.

Yuen, B., & Chin, H. C. (1998). Pedestrian streets in Singapore. *Transportation, 25*(3), 225–242.

Yuen, B. (2006). Reclaiming cultural heritage in Singapore. *Urban Affairs Review, 41*(6), 830–854.

Zmyślony, P., Kowalczyk-Anioł, J., & Dembińska, M. (2020). Deconstructing the overtourism-related social conflicts. *Sustainability, 12*(4). https://doi.org/10.3390/su12041695

Zuev, D., Psarikidou, K., & Popan, C. (Eds.). (2021). *Cycling societies: Innovations, inequalities and governance*. Routledge.

Index

Melissa Liow Li Sa and Sam Choon-Yin, *Sustainable Urban Development in Singapore*, Sustainable Development Goals Series,
https://doi.org/10.1007/978-981-99-5451-3

D

E

FSC
www.fsc.org
MIX
Papier | Fördert gute Waldnutzung
FSC® C083411

Kolibri 360 GmbH
Ferdinand-Jühlke-Straße 7
99095 Erfurt, Deutschland
produktsicherheit@kolibri360.de